C000176845

PUBLIC ORDER POLICING:

CONTEMPORARY PERSPECTIVES ON STRATEGY AND TACTICS

by Mike King and Nigel Brearley

Perpetuity Press

Perpetuity Press
PO Box 376 Leicester
LE2 3ZZ
Published 1996

Copyright © Perpetuity Press 1996

All rights reserved. Except for the quotation of short passages for the purposes of criticism and review, no part of this publication may be reproduced, stored in a retrieval system, or transmitted, in any form or by any means, electronic mechanical, photocopying, recording or otherwise, without the prior written permission of the publisher.

A catalogue reference for this book is available from the British Library

ISBN 1 899287 03 5 (paperback)

Printed by Biddles Short Run Books

King's Lynn

Perpetuity Press

Contents

Acknowledgements

Only through the contributions, suggestions and ideas of many academics and police practitioners is a work such as this possible. We would therefore like to thank all those who have taken part in this study, and more particularly our interview respondents who must remain anonymous. There is no way that a short book such as this can do justice to the wealth of information they supplied. We are especially grateful to Chief Inspector Chris Brightmore, Chief Inspector Lou Elliston, Chief Superintendent Michael Lofthouse, Chief Inspector Bill Morrell and Chief Inspector Matt Saunders, previous Public Order Masters students of ours who contributed their ideas to this study's formulation and arguments. Mention should also be made of some of the many individuals who have shared their ideas with us, assisted with archival materials and access or commented on various sections of this work, namely Chief Superintendent Doug Brand, Chief Superintendent Mike Davies, Inspector Richard Heritage, Chief Inspector John Lansley, Assistant Chief Constable Bill McGregor, QPM, Tony Moore, Dr David Waddington, Inspector Dennis Williams, Inspector Ron Woodham and Bob Wyllie, and all the participants of the Economic and Social Research Council's Research Seminar Series on the Policing of Public Disorder. Our thanks also go to Oliver Davies and David Hodgson for their archival research. Finally, we remain indebted to the Canadian Police College, Ottawa, for their sponsorship of our research.

Introduction

The image of the police officer in public order situations in England and Wales has undergone a profound change in the last 25 years. It has moved from that of the friendly, amiable 'bobby' to a 'militarised' and anonymous shield and baton-wielding, rubber bullet and CS Gas-deploying member of a Police Support Unit who wears flame-retardant clothing and a 'crash'-helmet with facial-contact obscuring visor and is perhaps even engaging in 'vehicle-tactics' against the crowd. Of course, the image does not necessarily reflect reality, as we suggest here. But it is undeniable that the form of policing has dramatically changed. Policing strategy and tactics do not operate in a vacuum, however; the process of change in them is subject to a number of factors at work both within and outside the police.

In this book we shall be attempting to catalogue and account for such change as has occurred within public order policing. We focus especially on the interaction between the police and the crowd, but do not totally ignore political and technological factors for example, although these are only indicated rather than developed here. The result of the process, we would argue, has been a heightened level of conflict and expectation of conflict. In this sense, the outcome at the peak has been a 'spiral of violence'. However, it has also given rise to an increased sensitivity on the part of the police, on the one hand to the potentiality for escalation in certain situations and on the other hand to the need for a more multi-agency approach to public order. In other words, it is not now simply a case of using the 'ultimate solution' first. Also, we draw attention in the conclusion to the current debate regarding changes in the nature of dissent, demonstration and protest in a period of significant social transformation variously termed 'late modernism', or 'postmodernity', which poses new challenges and questions about the possibility of exercising legitimate and consensual public order policing.

As one of our interview respondents suggests, contemporary public order policing resembles an iron fist in a velvet glove. The 'state of the art' wisdom would seem to be that the glove should not be removed until needs must, and then with the aim of containment, early resolution and post-event de-escalation. The problem, of course, in order first to maintain a situation of effective consensus policing and secondly to ensure that one's reaction does not cause an escalation is to match the strength of the 'fist' with that of the 'threat' *actually* posed. There has increasingly been a shift in public order policing away from the overtly confrontational situation *per se*, to the covert involving, as we shall suggest later, tension indicators, intelligence and evidence gathering, forward and contingency planning and consultation and negotiation. It is, however, necessary to stress here that the type of pre-emptive police response will also be determined by the type of public order situation which has arisen, as categorised *by* the police. A number of our respondents agreed with the categorisation of potential disorder situations as political, industrial, festival, urban and sport, and suggested that these would be policed differently unless a certain level of conflict ensued. We would add, however, that antecedent conditions may in fact cut across both the typology and the level of initial response. Further, as we stress in Chapter I, given our critique of the dominant perspectives of the crowd applied in public order training, as opposed to the 'rationality' of the crowd, the level of the initial police response generally has the effect of escalating or de-escalating potential conflict. We also argue, in Chapter II and in the Conclusion, that the changing nature of contemporary western society has given rise to a multitude of new issues, which is creating an increasingly complex mosaic of sites of legitimate protest which at least calls into question if it does not completely eliminate the traditional typology, and certainly has serious implications for the manner in which police forces plan to respond to potential disorder. At the time of writing, it would seem that the traditional typologies are still being employed by the police, despite their not fitting some of the present forms.

Our approach in writing this book has been to structure our considerations in three fairly distinct but overlapping strands. First, we offer a critical analysis of the way the crowd is represented in contemporary police public order training. Secondly, we undertake an empirical 'snapshot' review under a four-fold typology of the changing forms of policing and disorder from 1968 to the present. Thirdly, we consider the arguments for change in public order policing strategy and tactics in the light of a series of interviews conducted by us with senior police officers in England and Wales.[1] Finally, in the Conclusion, we draw together the main threads of our analysis and point to current indicators suggesting change in the direction of enhancing both proactive and reactive strategic and tactical forms of public order policing in association with the changing organisational form of the policing system generally. We also suggest the basis of current public order policing may well be compromised by the increasing tendency towards an apparent fragmentation of economic, social and ethical interests within contemporary society.

Glossary

Below we explain some of the terms employed in this book, especially in respect of the functions of agencies. We also give some indication of the economic, managerial and political changes that are occurring in policing in England and Wales and have an impact on public order policing. Terms in **bold** can be found either within the glossary or, where referring to case studies, in Chapter II.

1. We have included material from interviews undertaken by us and discussions with senior public order practitioners and trainers in England and Wales, including especially the British Transport Police, Kent Constabulary and the Metropolitan Police Service, from November 1992 until the present.

Association of Chief Police Officers (ACPO)

Although initially formed in 1948, since the 1964 Police Act the Association of Chief Police Officers (an essentially informal group representing officers above the rank of Chief Superintendent) has emerged as a pressure group with a strong influence on government policy-making and planning. Some **Chief Constables** speak out individually, but a doctrine of collective responsibility appears to discourage individual Chief Constables from criticising the ACPO 'line'. ACPO maintains a number of committees that gather information, lobby Parliament and the **Home Secretary** and provide advice and guidelines to the membership (Reiner, 1991).

There is an ACPO committee devoted to the issue of public order policing that has been highly influential in terms of training initiatives, the identification and exploration of tactical options and general policy development. The ACPO Public Order Committee prepares a Public Order Tactical Options Manual dealing with such issues as the nature and development of disorderly crowds and how they may be managed and controlled. This, in turn, shapes the format of local and central training documents and courses such as those developed by local forces and by the **Central Planning and Training Unit**. Such documents are not technically in the public domain. Under a legal ruling relating to a police force public order training manual, namely *Gill and another v. Chief Constable of Lancashire: Court of Appeal, 22 October 1992*, the manual is protected from disclosure at trial on the grounds of protecting the public interest unless the party seeking disclosure establishes that it cannot properly present its case without it. Northam (1988), however, does append extracts from the (then) ACPO Public Order Manual as deposited in the House of Commons library.

The Audit Commission

The Audit Commission is a government-appointed body charged with examining the efficiency and effectiveness of public and semi-public services. A number of Audit Commission reports have related to the provision of police services as part of a drive for efficiency and effectiveness, value for money and quality of service following a demand for improvements in these areas spurred by central government initiatives since the distribution of Home Office Circular 114/83, entitled *Manpower, Effectiveness and Efficiency in the Police Service*. These have had considerable impact on police organisation, management and internal funding arrangements and many more are promised in the future following the Sheehy Inquiry into Police Responsibilities and Rewards instigated by the **Home Secretary** which reported in 1993.

British Transport Police (BTP) and other statutory police forces

While there are currently 43 'local' or 'home' **police forces** in England and Wales, there also exist a number of other forces that have been created by legislation to police various specialist facilities. By far the largest of these forces is the British Transport Police, whose primary function is to police the railway network. Other statutory forces include those which police docks and military establishments. There have been efforts to privatise a number of specialised police functions, as well as to develop the use of locally funded officers with limited responsibilities. The privatisation of Britain's railway network has thrown further doubt on the continued existence of the BTP.

Central Planning and Training Unit (CPTU)

The CPTU is a centrally sponsored training establishment that runs and develops courses for all police forces, generally for officers up to and including the rank of inspector, notably in the area of training other officers.

Chief Constables and the Metropolitan Police Commissioner

Chief Constables are the heads of **police forces** in England and Wales. They exercise almost total control within their forces. Formally they are appointed by the **Police Authorities**, but the appointment is subject to approval by the **Home Secretary** who provides a list of suitable candidates. The Metropolitan Police Force has no Police Authority; the Home Secretary fulfils this function and appoints the Metropolitan Police Commissioner.

Criminal Justice and Public Order Act (1994)

This legislation sets out a range of offences, some of which are new whilst others which had previously been civil offences are moved into the ambit of the criminal law with the police being given the power to make arrests and pursue immediate resolution of contentious circumstances. Among the parts of the Act relevant to public order are the following:

Power of police stop and search (e.g. for use against suspected terrorists, but potentially much wider than this)
Section 60 grants police officers of superintendent rank, or above, the power to authorise uniformed officers to stop and search vehicles and persons within a specified area for offensive weapons or other dangerous articles. The authorising officer has to have reasonable grounds for believing that serious violence may take place in the area if such searches are not carried out. This power may be exercised for a 24 hour period and there is a provision for it to be extended for a further six hours.

Power to remove trespassers on land (e.g. for use against 'New Age Travellers')
Section 61 provides for a criminal offence of failing to leave land when ordered to do so by a police officer if that officer believes that two or more persons are trespassing, and if either damage has been caused or there are six or more vehicles on the land.

Powers in relation to raves
Section 63 gives the police powers to end preparations and gatherings for outdoor 'rave' festivals of 100 persons or more, or such festivals when in progress, at which amplified music is played at night and provides for a criminal offence of disobeying a direction to leave.

Section 64 provides the police with powers to enter land where an outdoor festival of the above kind is in preparation or progress and to seize vehicles and sound equipment.

Section 65 empowers the police to stop people whom they reasonably believe are going to an outdoor festival of the above kind anywhere within five miles of the boundary of the event, and direct them not to proceed.

Disruptive trespassers (e.g. for use against hunt saboteurs)
Section 68 limits the right to protest by creating a new criminal offence of aggravated trespass where a person trespasses on land in the open air and does anything intended to disrupt or obstruct lawful activity, or intimidate people so as to deter them from engaging in such activity.

Section 69 creates a new criminal offence where a person disobeys a police officer who directs them to leave land if the officer reasonably believes that person is committing, intends to commit, or has committed, an act of aggravated trespass. The offence occurs if the person concerned does not leave or enters again within three months.

Trespassory assemblies
Section 70 provides that an order can be obtained to ban assemblies which are likely to be held without the permission of the owner of the land or which may result in either serious disruption to the life of the community, or significant damage to a site of historical, architectural, archaeological or scientific importance.

Section 71 provides the police with powers to stop persons from proceeding whom they reasonably believe to be travelling to a prohibited trespassory assembly.

Offence of causing intentional harassment, alarm or distress
Section 154 creates a new offence under the **Public Order Act (1986)**. This offence of intentional harassment, alarm or distress which occurs when threatening, abusive or insulting words or behaviour are used or threatening, abusive or insulting writing, sign or visible representation is displayed, is linked to the Public Order Act offence of disorderly conduct and arises if the intent of the offender is to alarm other people.

Crown Prosecution Service (CPS)

The CPS was established in 1985 under the Prosecution of Offenders Act. It removes the burden of prosecuting cases off the police, although they prepare the case files. The CPS decides whether or not to proceed with prosecutions, thus introducing an extra level of accountability in the processing of cases. It is arguable that this has contributed to a 'tightening up' in the presentation of evidence and encouraged the development of detective work and evidence-gathering teams before, during and after public disorders. The CPS was established at a time of considerable change in the operation of the criminal law in England and Wales; other significant pieces of legislation include the Police and Criminal Evidence Act of 1984 (PACE), and the **Public Order Act (1986)**.

Gold, Silver, Bronze (GSB)

This is a set of command designations used in the policing of public order and based on military experience, which emerged from the major Public Order Review conducted after the **Broadwater Farm** and **Brixton** disturbances of 1985. **Gold** is the highest level of command and may be of any rank, depending on the predicted seriousness of the event. Gold, in consultation with superior and subordinate officers, sets budgets and staffing levels in advance of planned events and contingencies, and maintains overall control of an event from a command position, which is probably located at some distance from the event in an operations room. Essentially, Gold devises a plan to bring any incident to a 'satisfactory' conclusion. **Silver** command is generally in charge of officers 'on the ground' and issues major tactical instructions to the **Bronze** (sector) commanders in the event of unforeseen contingencies, whilst reporting to Gold. Bronze commanders issue instructions to officers on the ground, carrying out the instructions of Silver which are in accordance with the plan developed by Gold. Bronze commanders are typically the officers in charge of **Police Support Units**.

Her Majesty's Inspectorate of Constabulary (HMIs)

HMIs, established in their current form in 1964, are located in the **Home Office** and are responsible for inspecting **police forces** with a view to advising the **Home Secretary** on their efficiency. Their inspections rarely gave rise to contention until comparatively recently. One British police force, Derbyshire, twice recently had its certificate of efficiency withheld following local disagreements over budgeting. These inspections may have an impact on the local and central government funding of a force (Newburn, 1995).

Home Secretary/Home Office

The Home Secretary is the minister who heads the Home Office, the Department of State with responsibility for a whole range of national services including the police (and **Her Majesty's Inspectorate of Constabulary**) and fire services. There is much contemporary debate on the apparent increase in the centralisation of power over policing exercised by the Home Office at the expense of local scrutiny. The Home Office issues circulars offering advice on a whole range of policing issues to **Chief Constables** that rarely go unheeded.

Inner-city

This is an urban area typified by mixed usage, usually combining residential, commercial and recreational space. Multi-occupancy, low-quality, terraced housing is prevalent. Like many **peripheral estates**, inner city areas frequently accommodate those suffering from multiple deprivation in terms of housing, education and welfare services. Britain's black populations are disproportionately housed in inner city areas. Examples include **Brixton** in London, **St Pauls** in Bristol and **Toxteth** in Liverpool. Lord Scarman's report suggested that at the time of the **Brixton** disorders some 36% of the residents of the area were black (Scarman, 1982:22-29).

Inner-city housing estates

These are high density purpose-built unitary apartment complexes, usually developed as part of post-war slum-clearance programmes within inner-city areas, sometimes incorporating shopping precincts, public houses, schools and community facilities. They generally stand in marked contrast to the surrounding area. The **Broadwater Farm estate** in Tottenham, London, typifies an inner-city housing estate. This was completed in 1973 and consists of 1,063 apartments, mainly in 12 blocks each of between 4 and 18 floors, and a shopping precinct. It houses over 3,000 people at a density of population of 140 persons per acre. Broadwater Farm is constructed in such a way that car parking is at ground level and the apartment blocks are interconnected by walkways at first floor level. According to a survey conducted by the Gifford Inquiry in 1986, the adult population living on the estate was 49% white, 42% African-Caribbean, 3% South Asian and 6% other (Gifford, 1986:2.4-2.10, 2.28, 7.7). The majority of people on the Broadwater Farm estate live on very low incomes and in 1986 only 500-530 people of an adult population totalling about 1,800 were in full-time employment (Gifford, 1986: chapters 2 and 7).

Mutual Aid Co-ordination Centre (MACC)

Police Support Units from different forces can be co-ordinated and deployed in a crisis by what used to be known as the National Reporting Centre (NRC), now the Mutual Aid Co-ordination Centre. The NRC was established following the events at

Saltley Coke Depot during the **1972 Miners' Strike.** The Centre is located at the Metropolitan Police headquarters at New Scotland Yard and when in operation is controlled by the current president of **ACPO**. Mutual Aid arrangements were most prominently and controversially put into effect under the National Reporting Centre during the **Miners' Strike of 1984/5.**

National Criminal Intelligence Service (NCIS) and other central services

Established in 1992, the NCIS is a centrally-funded service designed to provide an information and service base for the investigation of crime. It is one of an increasing number of centrally provided services, and the policing structure in England and Wales is likely to undergo further centralisation, amalgamation and regionalisation. Other central services include forensic science laboratories, the Central Planning and Training Unit, the Police National Computer (PNC), the Bramshill Staff Training College (for senior officer training), as well as a number of departments at the Metropolitan Police headquarters at New Scotland Yard, such as the **Mutual Aid Co-ordination Centre** (formerly National Reporting Centre) and Metropolitan Area-specific support facilities, such as CO11 (formerly **TO20**).

Peripheral estates

These are usually post-war residential social-housing developments, such as **Meadow Well**, N. Shields, and **Wood End**, Coventry, which are often located on the periphery of large urban areas. They consist of 500 to over 2,000 houses and apartments, and accommodate mainly white working-class people. Like **inner-city** areas, peripheral estates have suffered from a lack of investment and many provide a bleak social and economic environment, often lacking in recreational and commercial facilities and opportunities.

Police Authorities

The primary task of Police Authorities is to 'secure the maintenance of an efficient and effective police service' in their local area. They have the power of scrutiny over local police spending (but not operational activities). Under the Police Act 1964 Police Authorities elected their own members being one-third local magistrates and two-thirds local councillors. With the Police and Magistrates Court Act (1994), however, there has been a shift towards increased central government control over selection and consequently public spending. Police Authorities may now elect their own chairperson and a local selection panel will be established to select a number of non-elected members. This selection panel consists of a Home Secretary's appointee, a member selected by the Police Authority and one selected by both. The panel draws up a short-list of candidates which is then subject to further scrutiny by the Home Secretary. From this, the magistrate and councillor Police Authority members select the successful candidates. The Metropolitan Police Service does not have a Police Authority, but rather is directly subject to the Home Secretary (Loveday, 1995; Newburn, 1995).

Police forces

After the implementation of the 1964 Police Act, numerous amalgamations of **police forces** occurred, eventually resulting in the 43 'local' or 'home' police forces that currently exist in England and Wales. These vary considerably in size and may cover a county, for example, Kent, Essex, Lancashire, a group of counties, for example West Mercia, South Wales, or a major metropolitan area, for example the (London) Metropolitan Police Service (in addition to the City of London Police) and the Greater Manchester and Merseyside Police Forces. A major review of policing was instigated by the **Home Secretary** in 1992 under the chair of Sir Patrick Sheehy. His committee reported in 1993 and, along with many other recommendations, suggested some further amalgamation of forces with a view to improving communications, efficiency and effectiveness and liaison with central services. The Police and Magistrates Court Act (1994) provided for further amalgamations as well as affecting the extent of police accountability to locally elected and appointed **Police Authorities.**

Police Support Units (PSUs)

These are tactical units, specially trained and equipped for deployment in public disorder situations. They had their origins in 1965 with the **Special Patrol Group** (SPG) in the Metropolitan Police (since disbanded, following especially the death of Blair Peach at Southall, west London in 1979 and the subsequent inquiries (P. Waddington, 1994:18)rr. Variations exist such as Territorial Support Units (TSGs), and the Mobile Support Units (MSUs) in the **British Transport Police**. PSUs were restructured in the early 1990s to ensure closer supervision. Each **police force** is required by the **Home Office** to train and maintain a given number of PSUs for deployment under Mutual Aid arrangements.

Public Order Act (1986)

The 1986 Public Order Act largely replaced the provisions of the 1936 Public Order Act; it codified and strengthened the law in this area, redefining many common law offences and incorporating them into one coherent piece of legislation. The scope of the Act is such that it is used to prosecute a wide variety of public order offences, ranging from altercations and fights in the street, through to full-scale riots. It also contains important provisions regarding picketing, and the banning and prior notice of demonstrations and meetings. The Public Order powers of the police were considerably extended by the **Criminal Justice and Public Order Act (1994)**.

Territorial Operations 20 (TO20)

Until April 1995 TO20 was the central public order intelligence, planning and operations section within the Territorial Operations Branch of the Metropolitan Police Service at New Scotland Yard. With the reorganisation of the Metropolitan Police into five areas, each under an Assistant Commissioner and with its own matrix responsibility, public order responsibilities have by and large been dispersed to the individual areas, although Central Area has retained the matrix responsibility for this. **Commissioner's Office 11 (CO11)** has taken over the remaining public order tasks from TO20 including public order intelligence (although **NCIS** has some public order intelligence responsibilities also).

CHAPTER I
The Disorderly Crowd: a critical analysis

Introduction

In this chapter we offer a critical analysis of work on the nature of the crowd which has some bearing on the development of techniques of control that may be deployed by police organisations. The research for this chapter was primarily conducted at the Home Office Emergency Planning College.

We first comment on the models of riots and categories of crowds that are commonly employed to promote awareness and discussion of public order issues on police training courses at a variety of levels. We will focus on the adequacy of the 'riot curve' and a model of crowd classification. Secondly, we examine critically the work of Neil Smelser, a writer whose work commonly influences the development of training packages. We suggest that Smelser's work, while avoiding some of the theoretical excesses of earlier writers who regarded crowds as somewhat pathological entities, still fails to take into account the essential rationality of much crowd behaviour and the possible contribution of police action to the development and activity of crowds. Each stage of Smelser's theory will be criticised by reference to subsequent writers who, in general, adopt approaches which show greater awareness of the rationality of crowd participants and of the impact of control practices on their behaviour. This critical exploration will provide the basis for comment on the implications of these issues for policing policy and practice. We argue that a full appreciation of the social structural context of crowds requires the acceptance of a model of human activity within crowds that acknowledges the rationality of participants. This, we suggest, provides the most appropriate basis on which to build an effective policy for public order policing.

The riot curve

We are informed that there is a tendency in a number of the police public order training courses in England and Wales, for reference to be made to a 'riot curve'. The precise form of this curve varies, but the example that follows may be regarded as typical:

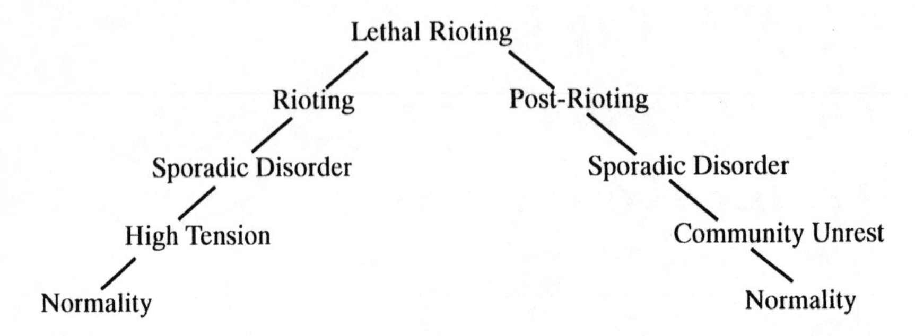

The influential police commentator Beckett (1992) employs a similar approach in his exploration of current public order policing strategies in England and Wales. Beckett's description may be expressed diagrammatically as follows:

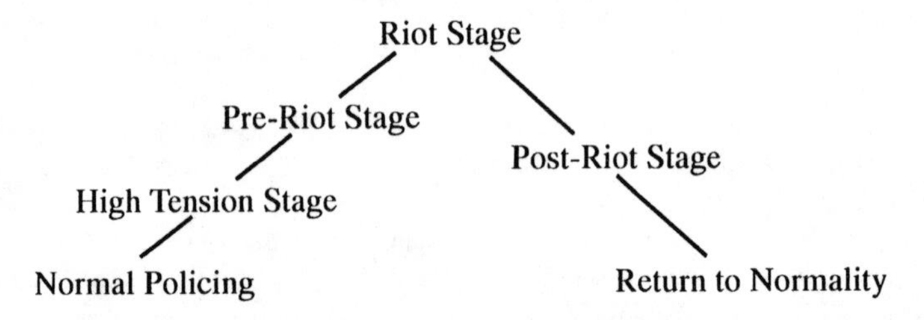

Both of these models present public disorder escalation as being a dynamic process, initiated by the public, as crowd participants, and responded to by the police. This is an implication which we will dispute, arguing instead that police forces and their activities are a part of an interactive process involving the crowd and the public and that police preparations for disorder, and their responses, constitute essential ingredients in the development of disorders.

Beckett suggests that in ordinary circumstances the primary aim of policing is preventive and involves 'employing community mediation services to resolve isolated outbreaks' of trouble. In the high-tension and pre-riot stages, for example, 'conflict reduction (or de-escalation of the temper of relationships) would be the dominant aim, employing joint police and community action'. However, in the event of a full-scale riot developing, Beckett argues that the police would only have the minimisation of harm and damage to attend to as options for resolution would have been exhausted and the time would hardly be ripe for attending to the underlying problems (Marshall, 1992:91).

In managing the early stages of potential disorder, Beckett advocates a 'contact strategy'. This approach emphasises the development of flexible relationships with local agencies and communities and stresses the preventive potential of 'normal' policing practices.

Beckett's view, however, remains firmly focused on police action responding to that of the public. He argues that the development of conflict demands that the degree of force employed by the police should *follow* and *match*, but not *exceed* the level of antagonism shown by disorderly elements. In this way Beckett argues that police action should correspond to the prevailing mood, but not provoke an escalation of violence. However, as Marshall (1992:91) notes:

> One might argue that there are problems in this - that the police are in danger of losing control of their own actions, which are determined precisely by the initiatives taken by other groups. It is possible to argue that in some cases the police should be prepared to show more restraint than others, not to respond in the same measure to increases of aggression in others, in an active attempt to de-escalate.

Marshall goes on to stress that Beckett's approach is openly and appropriately that of a police officer and notes the misunderstandings that this may give rise to among observers and commentators:

> His definition of a riot is purely in criminal legal terms: it has to involve general looting or an attack upon the police. In lay eyes, a riot may not have quite the same connotations. This could lead to misunderstandings. When the police see a change from one "stage" of conflict to another, community members may not see it in the same way, seeing the situation as more or less severe. This could lead to conflict over changes in police operations that are seen as unwarranted or charges of inaction when intervention seems to be called for but does not occur (Marshall, 1992:91-2).

Further to such criticisms, recent experience in England and Wales suggests that these models are not always adequate for police purposes and can not be fully relied upon for effective public order planning, whether for particular events or contingencies. Whereas one incident may go from High Tension through Sporadic Disorder, Rioting, Serious Rioting and on to Lethal Rioting, another may travel from High Tension to Serious or Lethal Rioting very quickly, by-passing the gradual build-up. Judgement of the local potential for disorder to develop and risk assessment of the development of disorder in particular circumstances is an aspect to which we shall return in Chapter III when considering the development of Tension Indicators and the interpretation that is applied to them.

How people behave in crowds, and how far being part of a crowd may lead to changes in people's behaviour has been a matter of considerable debate for more than a century. We intend to explore some of these issues and the conclusions may help to explain why the riot curve is perhaps an inadequate model, as it appears to assume a 'natural' escalation of violence on the part of a crowd, regardless of the environment and police action and the response to it. We will therefore continue our examination of the types and nature of crowds, first by examining a model of crowd categories and the relationships between them; and secondly by examining theories of crowd behaviour.

Types and categories of crowd

Many types of crowd present few problems for public order and there is a temptation for policing purposes to divide them into two categories: 'Audiences' and 'Mobs'. This categorisation is based on information received from serving police officers in England and Wales, and is taken as being representative of the ideas presented in police training at various command levels. Such training, while varying in precise detail, generally entails crowd categories being presented in the following manner:

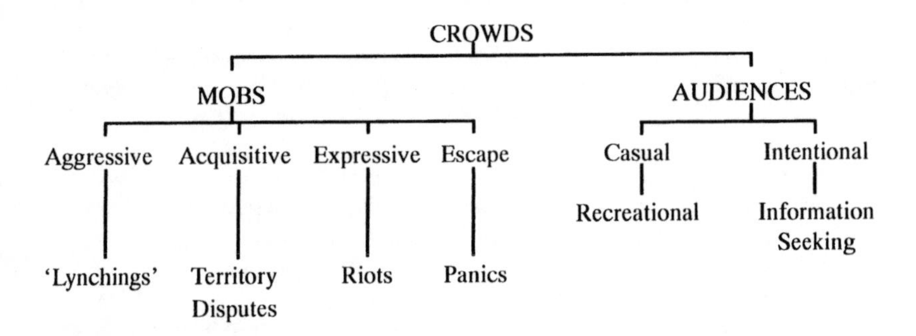

The problem with such a model is that some of these categories have a tendency to flow into one another over time in a manner that is influenced by a number of factors, including the geography of the scene and the response of those with responsibility for control. In this sense it only gives a limited picture of the dynamic and potentially changeable nature of crowds. For example, the witnessing of a police action such as an arrest, or even helpful behaviour as in the case of the disturbance that occurred in Brixton in 1981, can lead a casual audience to cross the boundary into 'mob' activity. In the Brixton case, this 'mob' activity included a variety of aggressive/expressive territorial and riot responses and was accompanied by acquisitive behaviour. Similarly, a protest demonstration, which may effectively constitute an information-seeking crowd with expressive intent, may, if inappropriately organised and handled by the police, perhaps being directed into a restricted space, be transformed into a 'mob'. In response to circumstances it may change into a crowd that, in a panic to escape, does damage and provides the occasion for riotous behaviour, aimed aggressively/defensively at those directing the participants into an environment that they perceive as dangerous, or otherwise undesirable.

This simple classification above fails therefore to accommodate the sort of dynamic shifts from one to another of its categories that may be either the product of changes in the internal nature of the crowd, or a response to police activity designed to impact on the crowd. In this sense, such a model does not fully reflect the experience of many of those who have participated in crowd events and/or had responsibility for managing and controlling their evolution.

Causes of crowd behaviour

Thus it is necessary to examine crowd dynamics and the circumstances surrounding the relationship between the crowd and the agents responsible for its management and control. Since disorderly events occur in what is frequently complex urban and social space, it is necessary to be aware of a range of considerations drawn from the social and behavioural sciences (Gaskell and Benewick, 1987; Dunning *et al.*, 1987; D. Waddington *et al.*, 1989; D. Waddington, 1992; Marshall, 1992).

Early efforts to account for crowd behaviour concentrated on the psychology of crowd participants and, as with early criminologies, tended to see disorderly crowd participants as pathologically influenced, their actions being devoid of genuine social meaning and individual intention. The first influential writer on the crowd was Gustave Le Bon who saw the crowd as irrational, degenerate and pathological. Le Bon explained the apparent co-ordination and co-operation in crowds as the product of a 'group mind', a term that is still used today, although with rather less emphasis. This group mind emerged through a process of contagion affecting those made susceptible by the submergence of individual personalities in the will of the 'libidinous' mob. The lack of individual control and rationality exhibited by crowd participants suggested by Le Bon is starkly described in his classic and influential book *The Crowd,* first published in 1895:

> ... by the mere fact that he forms part of an organised crowd a man descends several rungs in the ladder of civilisation. Isolated he may be a cultivated individual, in a crowd he is a barbarian - that is, a creature acting by instinct. He possesses the spontaneity, the violence, the ferocity and also the enthusiasm and heroism of primitive beings, whom he further tends to resemble by the facility with which he allows himself to be impressed by word and images (Le Bon 1960:32-3, cited in Gaskell and Benewick, 1987:2).

This perspective has been further developed and remains influential:

> Allport (1924), developed the theme of contagion by proposing a circular chain reaction of emotional facilitation. Individuals respond to and stimulate others generating an ascending spiral of crowd emotionality. Recent work on crowd facilitation (Zajonic, 1965) has been cited to support this hypothesis. The theory of deindividuation (Zimbardo, 1969) is a variation on the same general theme. Here the mechanisms underlying aggressive and uninhibited behaviours in crowds are anonymity, arousal, similarity of dress, unstructured situations and the very involvement in an aggressive act. A large, densely packed and noisy crowd leads to heightened arousal of the individual,

> enhancing the slide from autonomy and restraint to deindividuation, a state of diminished self-awareness and self-control. Such individualistic theories appear to explain the homogeneity of the crowd but do so at the cost of predicting bedlam in every crowd, at Ascot and the lynch mob alike (Gaskell and Benewick, 1987:2).

It is understandable that the concept of a 'group mind' is a tempting one for officers charged with public order maintenance to adopt when faced by a hostile crowd venting its anger but, as we shall see, it is simplistic and ultimately unhelpful for the development of effective and legitimate public order policing.

More recently, efforts have been made to classify crowds and explain their behaviour in broader social terms, moderating the unduly mechanistic prescriptions of writers such as Le Bon by incorporating an understanding of the social structural origins of the crowd and the meanings attached to their actions by participants. The historians Rudé (1964) and Hobsbawm (1959), for example, have done much to generate awareness of the rationality of crowd participants in history. Subsequent studies have explored the social background and investigated the motivations of participants in disorderly crowds. Work such as that of Sears and McConahay (1973) and Feagin and Hahn (1973) explored the backgrounds, motivations and activities of riot participants in the United States, and Cooper (1985) investigated the backgrounds of those arrested in the Toxteth riots of 1981 (Gaskell and Benewick, 1987:2-7). These studies revealed that crowd participants are cognizant of the 'appropriateness' of what they do and are frequently concerned to legitimise their participation and justify their actions. But this would not appear always to be the case. As Gaskell and Benewick point out:

> Some have used the evidence of 'representativeness', the 'post-hoc' labelling of behaviour in terms of protests, and of flashpoints involving the police as justifications of riotous behaviour. This is as misleading as the opposing tendency to dismiss the rioters as 'riff-raff', 'deviant' or 'criminal', and their actions as mindless response to 'outside agitators'. Theoretical work attempting to understand the crowd in terms of a multi-disciplinary social scientific perspective may develop a deeper understanding (Gaskell and Benewick, 1987:4).

An attempt to achieve a multi-disciplinary social scientific approach was made by the American sociologist Neil Smelser. He set out to merge the valuable, albeit previously overstated, contributions of psychology with a broader understanding of the social processes that shape individual actions within crowds. We have chosen to employ Smelser's work as an initial focus, not because we are fully supportive of his explanations or prescriptions, but because our enquiries among serving police officers in England and Wales suggest that his perspective has been particularly influential in shaping current views on the development of public order policing.

While Smelser's work is grounded in structural functionalism and provides a general model of social action, what concerns us here is his specific model of collective behaviour. Smelser's account of the causes of crowd behaviour can be set out under six broad headings:

i. Structural conduciveness
ii. Structural strain
iii. Crystallization of belief
iv. Precipitating factors
v. Mobilisation of participants
vi. Social control.

We will examine each of these levels of explanation and put forward our observations on their implications for policing practice on the basis of subsequent studies and commentaries. This section will provide us with a set of theoretical propositions on which to base an examination of specific instances of public disorder in England and Wales in Chapter II. We acknowledge that Smelser's account of disorderly crowd causation has made a major contribution to broadening awareness of the factors that may influence crowd behaviour and represents a distinct improvement on earlier mechanistic psychological models. We argue, however, that it remains somewhat limited, like the training models that we have examined, in that it fails to take sufficient account of both the generally rational nature of crowd participant behaviour and the possible impact of the complexities of macro-level social conflict and diversity on policing policy in relation to specific events. In this sense we are in agreement with the analysis of Smelser's work by D. Waddington *et al.*, who note:

> Smelser's commitment to the construction of a general theory of collective behaviour ... (is) meaningful only to those who see society as a self-regulating system ... Underlying roots of conflicts are taken to be less important than the characteristics of the situation in which potentially conflicting groups find themselves ... This is further evident in the little emphasis placed by Smelser on the historical and political contexts in which disorder may arise (1989:174).

i. Structural conduciveness

This refers to the general social conditions that are necessary for the development of a 'collective episode'. D. Waddington *et al.* define structural conduciveness as: 'physical and organisational facilitators/constraints, differing according to types of behaviour' (1989:173). Generally, for a disorderly collective episode to occur the environment must contain: an agency to which the blame for an unsatisfactory state of affairs may be attached; an absence, or failure, of channels to manage grievances; and means for the aggrieved to communicate and come together.

It is argued that these conditions are likely to be met when the environment fails to provide other channels of dispute resolution or methods of protest and change. Comments by politicians and/or community leaders may assist in the production of a social climate in which crowd action potentially prejudicial to established order is encouraged, or at least tolerated. An aspect of structural conduciveness of particular relevance to urban disorder is economic stringency, which may generate social frustration producing aggression that may emerge in many ways, including expressive and acquisitive forms of public disorder. It is also likely that the police will be the social agency immediately present at the scene of potential disorder and thus provide a focus for resentment or else be charged with interposing themselves between the aggrieved and their intended targets.

In democratic societies, the police role in preventing the development of disorderly structural cohesiveness is somewhat limited and mainly consists in communicating to politicians and community leaders the possible consequences of their actions. Such communication may take the form of comments on the impact of government economic policy, a comparatively rare but increasingly frequent occurrence in England and Wales; or advice to community leaders on the possible impact of their actions on the community in terms of public order policing and encouraging negotiation and the organisation of peaceable protest.

Despite their lack of any formal role in this field, this is an area of great interest to officers with responsibility for forward planning for the ventilation of dissent and the control of that which occurs spontaneously. It is therefore a vital element in the monitoring of community tension. For ground commanders such awareness is also vital if they are to use their resources in such a way as to present information to crowds, exercise control over space and lines of communication, and act in such a way as to encourage the re-establishment of peaceable community relations. As we shall emphasise, it is likely that this field of policing endeavour will be of increasing importance given the diversification of protest in a society of fragmenting interests. In this context the idea of 'community' should not necessarily be seen as a geographically bounded concept, although the experience of inner-city and peripheral estate disorders combined with the local orientation of policing in England and Wales means that the term is much used in this sense. Personal mobility, the rapidity and extent of media dissemination of information and ideas, and the changing nature of the polity, are all increasing the importance of communities of interest, or lifestyle, as distinct from occupational or residential locality.

ii. Structural strain

This term is used to describe conditions where aspects of the social, economic and political system are 'out of joint', giving rise to sources of conflict for which there is no immediate solution or mitigation and thus to structural conduciveness as outlined above. The strain may arise from group antagonisms based on economic, political, religious and/or cultural divisions. It may also arise from a group's feelings of frustration in the face of rising material and social expectations, or economic conditions leading to a decline in material and social expectations. Waddington *et al.* define this aspect of

Smelser's model thus: 'conflicts of interest; means-ends ambiguities; normative malintegration; differences in values' (1989:173).

As with the conditions of structural conduciveness outlined above, there is not a great deal that the police *per se* can do about these structural conditions, other than to contribute informed advice to the political debate, make efforts to minimise the effect of such conditions on the policing of the community and provide protection against conventional crime. Since structural strain leads to conflict over the allocation of wealth and resources in communities, and it is likely that aggressive and acquisitive solutions will be sought, it appears inevitable that the efforts of the police to prevent crime will raise tensions and lead to allegations of partisan policing, which may make the police a target for dissatisfied crowds.

In addressing the problems posed by structural conduciveness for the development of a disorderly crowd and for criminal activity, the police will find themselves in an unenviable and delicate position if they are to maintain the peace and protect some elements of the community against others, while minimising the strain between themselves and those whom they aim to police by consent. The development of police understanding of tensions within local communities and of the potential threat posed by the demands of interest groups plus reasoned and, if appropriate, sympathetic debate on the problems to which some may see disorderly conduct as being an immediate and appropriate solution, appears to be the best course to follow in terms of effective long-term policing practice.

iii. Crystallisation of belief

This is an area of theories of crowds of which we are highly critical, regarding the use of this terminology as something of a throwback to early mechanistic psychological explanations of behaviour, according to which crowd participants lose all free will and become tools in the hands of 'political extremists', or mere products of base human instincts as described by Le Bon. Smelser's work is somewhat sophisticated in that it does appear to allow for a limited degree of identification by crowd participants. Smelser argues that structural conduciveness and strain produce generalised beliefs within groups, giving rise to a common appreciation of their social circumstances - a 'crystallisation of belief' - which may take a form hostile to other groups or agencies.

Smelser identifies five stages through which hostile beliefs may crystallise: a state of ambiguity emerges from social strains; tensions and anxieties mount; the aggrieved communicate their tensions and fears amongst themselves and allocate blame to the social agents that they have identified as responsible for their position; a desire to punish those whom they see as responsible for their grievances, or those protecting these social agents, develops; and, finally, through the experience of collective grievance and purpose, a generalised belief in their omnipotence emerges.

In this sense the behaviour of crowd participants may well be shaped by the structural circumstances that bring them to be participants in the crowd, and they may well behave in a manner that is different, perhaps more committed and extreme, than if they were acting as individuals or in small groups; but this is not to say that they lose all reason. We would argue that participation in a crowd may give individuals the confidence to behave in more extreme or uninhibited ways than usual. Participants may take more risks since they are not alone, or doubt the likelihood of their being called to account for their actions but, we would argue, they do 'know what they are doing'. Indeed, it is interesting to note how many conventional commentators on the political right are tempted to adopt a 'group-mind' interpretation of public disorder, referring to 'riff-raff elements' and acts of 'mindless hooliganism'.

We were informed by police trainers that early efforts to employ crowd theories in basic training have tended to focus on the later stages of the development of hostile belief at the expense of developing an understanding of the social structural and community factors. Concepts such as 'contagion', 'modelling', 'invincibility' and 'anonymity' are stressed in developing officers' appreciation of the nature of crowds. We comment on the value of these ideas below. Our general line of argument, however, is that such concepts convey an excessively mechanistic view of the actions of crowd participants and that an over-concentration on these aspects may lead to the potential for intervening in the antecedent conditions, and so heading off disorder, being neglected:

Contagion - the spread of behaviour from one person to another. The facial expressions and gestures of one person induce others, or, in our appreciation, give others the confidence to express similar emotions. We doubt that contagion is a mechanistic process and would argue that the broader social structural appreciation needs to be shared before the individual takes on the emotional response of another.

Modelling - this is related to the concept of suggestibility, but does not have the same mechanistic psychological implications. Often we behave in the way that we do because we see others that we respect or wish to emulate behaving in a similar way. If salient figures in a crowd, often crowd leaders, start to act in an illegal or dangerous manner, we may expect others to copy them. However, we would argue that while there may be exceptional cases where the uninitiated get 'carried away', those who are most likely to emulate such forms of behaviour are individuals who share, to some extent, the opinions shaped by social structural factors and/or the experiences of the instigators. The extent to which others will choose to emulate disorderly leaders can be reduced by drawing attention to those who are behaving in a law-abiding way.

Invincibility - feelings of invincibility would, in our terms, be defined as the development of vastly enhanced levels of personal confidence in one's capacity to promote interests and express opinions by disorderly means, gained from participation in a crowd. In Smelser's work it is argued that the seething and milling of the crowd may itself contribute to an individual's sense that the crowd is all-powerful and that normal social rules can be abandoned.

According to Smelser, the way in which a crowd comes together is of vital importance in determining its structure. A crowd must reach a certain size, or 'critical mass', in order to generate the milling and the anonymity which help to generate a 'group mind' or identity - all-important factors in creating threats to the preservation of public order. Smelser notes that within crowds it is often a small group that is the focus of serious disorder and how this group is dealt with may have substantial impact on the behaviour of many more. Estimates of crowd size are notoriously inaccurate, not only by police representatives, but also those made by organisers and other, supposedly knowledgeable, participants, making a strict rule about crowd size and behaviour impossible to formulate, even if the 'quality' of the crowd and its *raison d'être* are taken into account.

While increases in crowd size and density may pose problems for public order and safety, the dispersal of a crowd may also be a period when considerable crime and disorder occur. This is evidenced by a number of the Notting Hill Carnivals in London, to which we refer in Chapter II. In such cases street robbers have preyed on revellers making their way from the carnival by public transport as they became progressively detached from the celebratory and supportive mass of carnival participants; and also when a distillation effect has occurred, with law-abiding participants quitting the event at around the time of its negotiated closure, leaving a critical mass of highly involved people to oppose the police presence and resist efforts to close the celebrations (Roach, 1992).

Crowds tend to develop in a circular manner, sometimes affected by buildings or other immovable features. The most active participants are usually at the centre of the circle, and similarly if the crowd is, or becomes, disorderly, it is likely that the inflammatory elements will be near there also (Lewis, 1992). Should the police choose to enforce the law against the illegally active members of a crowd there is a very good chance that they will have to move through the law-abiding participants to reach the more militant elements. If this is not carried out in a courteous way, the law abiding may well become alienated and actively inhibit or counter police action, escalating the level of disorder.

'Laminated crowds', embodying strata of people representing opposing interests, are likely sites of disorder and should be avoided by planning, negotiation and management. These used to be common in England and Wales especially in cases of political protest, for example when the National Front, and related interests, staged controversial marches that drew opposition gatherings in the late 1970s, for example at Lewisham in 1977 (as noted in Chapter II) and at Southall, west London in 1979 (NCCL, 1980) where an anti-fascist demonstrator, Blair Peach, was killed, apparently from a blow to the head by a member of the SPG during a baton-charge (Rollo, 1980:158; P. Waddington, 1994:18). Improved negotiation, management and organisation, facilitated by forward planning and the application of public order legislation, have done much to avoid these forms of confrontation, although they do still occur, as in the case of the clashes between the British National Party and counter-demonstrators at Welling, south London in 1992 (P. Waddington, 1994:167-171) and in 1993 (D. Campbell, 1993).

Boundaries of crowds vary in their permeability and definition and much public order police work involves negotiating these boundaries. Depending on the state of the law, a crowd may have to be penetrated to facilitate the lawful entry of celebrities or officials, workers choosing to cross a picket line or even, as we are currently witnessing in England and Wales, the export of live animals for slaughter and calves for incarceration in veal crates abroad (RSPCA, 1995). In some circumstances it is necessary for the police to create a boundary that is permeable in one direction only. When containment is exercised to permit only egress from a crowd, this may have the advantage of facilitating the departure of participants who wish to leave a crowd that they perceive to be getting out of hand, while discouraging the addition to the crowd of others who may wish to contribute to disorderly behaviour.

Anonymity - feelings of anonymity are also said to be at the core of much anti-social crowd behaviour. It is difficult to locate and apprehend people in crowds and it is this knowledge, backed by the confidence to engage in anti-social and illegal acts resulting from the factors outlined above, that may give potential offenders the confidence to act.

Efforts to control crowds by formalising their seating or standing arrangements may bring about a stabilisation of their characteristic milling movement. The recruitment of responsible civilian stewards by event organisers may also have a positive impact, especially if they are well supervised and trained to direct crowds politely and facilitate their lawful activities. Sadler's research on public order policing practices in Europe noted that police forces recognised that stewards were valuable: 'to keep demonstrations on the route agreed with the police; to defuse trouble spots by isolating trouble-makers or agitators and dealing with them non-violently; (and) to act as "go-between" or communication link between police and demonstrators'. In the Metropolitan Police Service, she found that it was common for stewards to facilitate police action against unwanted activists and that this was appreciated, although some officers rather felt that stewards might be usurping their role (Sadler, 1992:125-126).

It is also important for the police, in collaboration with organisers, to make appeals stressing individual accountability. For policing purposes it would appear that the risk of trouble may be reduced by making sure that crowd members are not crushed together and have the choice of moving away from potentially contagious, or in our less mechanistic conception, 'inspiring' circumstances. A complementary tactic that may be deployed is that of drawing attention to peaceable elements in the crowd and facilitating their law-abiding behaviour, while distancing their activity from that of those posing a public order problem.

iv. Precipitating factors

The previous three stages of Smelser's model merely set the scene in which violence and disorder may occur. In many cases these circumstances are not enough in themselves to generate disorderly behaviour on the part of the majority of the crowd. While structural conduciveness and strains may give rise to a crowd that presents problems for the maintenance of order, and provide an environment in which confidence is gained for the commission of criminal acts, precipitating factors are necessary to provide a focus and perhaps legitimacy for such activity. These may be many and varied. D. Waddington *et al.*. summarize precipitating factors as '... incident or rumour of incident indicative of breakdown of relations; hostile outburst; sudden threat or deprivation' (1989:173).

In a number of cases we have seen minor traffic violations, or the mere observance of an incident, become the occasion for crowds to turn from relatively peaceable protest, into mobs. In the case of the Brixton disorders of 1981, intensive police activity in the area had aroused animosity, but it was the local misinterpretation of the rendering of help by the police in the case of a street attack that provided the precipitating factor. At Bristol in 1980 a police raid on what was suspected to be an unlicenced drinking establishment sparked an extreme community response. In Brixton in 1985, and in Tottenham in the same year, police raids leading to a death and serious injury to women who were not targets of the police operation, similarly provided precipitating factors.

Such factors need not even be as tangible as in these examples. Rumour, true or otherwise, may be sufficient to precipitate disorderly crowd behaviour. It was long believed that these rumours were generated by individuals who wished them to be true. This 'wish fulfilment' theory has more recently been challenged and largely replaced by the idea that when people feel angry, anxious or frightened a rumour can sometimes make them feel better by providing a justification for their emotions (Rosnow, 1980; Colman, 1991a,1991b). Smelser too regards rumour as being important in the generation and crystallization of hostile belief. It is further argued that, while rumour may constitute a precipitating factor in itself, other factors, such as varied and uninformed interpretations of a police action, may have an effect on the growth of rumour and belief. Such precipitating factors may confirm existing hatreds, introduce new forms of deprivation, reduce opportunities for peaceful protest and symbolise a failure for which responsibility must be assigned (Smelser, 1962:249-251). We would stress, however, that what we have described are *precipitating factors*; they do not explain the occurrence of riot in themselves. Any major disorder that ensues has a multitude of structural causes that precede the event and make the precipitating factor highly significant to those who participate in the crowd. We would argue that 'precipitating factors' are perhaps better understood in relation to the 'flashpoints' model of disorderly crowd causation elaborated by D. Waddington *et al.*. (1989) and D. Waddington (1992) and referred to in Chapter II. One faces a major dilemma in assessing what may constitute a precipitating factor to riot and what may be a deterrent. For example a strong police presence, overt or covert, may be just as likely to be the precipitating factor for a riot as to act as a deterrent, and contingency planning for either eventuality is increasingly being regarded as an important aspect of police competence.

In the case of rumours, with their roots in fear, anxiety and anger, that may serve to escalate such emotions and fuel a potentially disorderly situation, it would appear that they are best countered by calming emotions and by providing as much information and evidence as possible. This suggestion is based on psychological research into the development of rumour, but it must be acknowledged that the emotions that are the seedbeds of rumour may well have been aroused before the potentially disorderly event. For escalatory rumour to be effectively countered, those providing information designed to calm and de-escalate a situation must be credible commentators in the eyes of those whom they are addressing. This can only be achieved by developing an atmosphere of mutual trust and respect in advance. Thus there is a need for the police to develop and maintain strong and positive links with the communities they are policing.

v. Mobilisation of participants

Smelser points out that crowds which are likely to engage in disorderly behaviour must contain reasonably competent and knowledgeable leaders and that these leaders must have some means of communicating with the crowd. For disorderly behaviour to be maintained over any given period there must also be means of communicating with the wider community in order to attract reinforcements. Again, the transmission and credibility of rumour is an important factor here. However, the influence of leaders and their potential to persuade others to follow their behaviour, by direction or by example, may frequently be overestimated, notably by the popular media. Lewis (1992) notes that crowds are complex social systems, and suggests that participants may be categorised according to three different roles, namely the active core, cheerleaders and observers

The active core carries out the action by which the crowd is frequently typified, despite being in a minority. Its members may be regarded as the real protagonists. The cheerleaders act in verbal or symbolic support of the active core, applauding and supporting its activities. When the core is active, the cheerleaders may 'hang around' or rush about and perhaps even take the opportunity to loot, while the core confronts the police. Lastly, any crowd will contain a number of observers who follow the actions of the active core and the cheerleaders, but seldom, unless mobilised by extreme precipitating factors, take part in their activities.

vi. Social control

We would regard this as one of the weakest aspects of Smelser's model in terms of its explanatory value and in terms of the prescriptions that he derives as far as policing practice is concerned. It would appear that Smelser advocates a 'short, sharp shock' approach to the control of disorderly crowds (D. Waddington *et al.*., 1989:174-175), with little regard for the consequences. In Smelser's own words: 'When the authorities issue firm, unyielding and unbiased decisions in short order, the hostile outburst is dampened' (Smelser, 1962:265, cited in D. Waddington *et al.*. 1989:175).

The theory, following from Smelser's exploration of the role of leaders, stresses the need to interrupt the flow of communication between those seeking to provoke disorder and the rest of the crowd. Perhaps more seriously, Smelser also implies that it is desirable for the police to avoid developing a sensitivity towards the mood of different elements in the crowd. His suggestions are as follows:

> a) Prevent communication in general, so that beliefs cannot be disseminated. b) Prevent interaction between leaders and followers, so that mobilisation is difficult. c) refrain from taking a conditional attitude towards violence, by bluffing or vacillating in the use of ultimate weapons or force. d) Refrain from entering the issues and controversies that move the crowd; remain impartial, unyielding and fixed on the principle of maintaining law and order (Smelser, 1962:267, cited in D. Waddington *et al.*. 1989:175).

Smelser does not, however, explain how interaction between leaders and followers is to be prevented, and focuses heavily on the police (or the military) as the sole agents capable of calming disorder. On the basis of the review of recent experience of public disorder in England and Wales contained in the following chapter we agree with D. Waddington *et al.*. that Smelser's 'prescriptions for avoiding disorder seem ... more likely to cause it' (1989:175).

In any of the six situations covered by Smelser which may lead to the development of a disorderly crowd, various measures may be taken by a range of agents of social control to reduce the possibility of a riot developing. The police are by no means the only agents of social control; community leaders, local and central government politicians, interest groups, the media, the organisers of marches and events; families, friends and other elements of the community can all play a part in preventing the development of serious conflict. It is essential that police organisations bear this in mind when planning their response to circumstances that may generate crowds. It is clear from the comments of Beckett on 'contact strategies' that this is now well understood by many in the police (Beckett, 1992: 133). However, marshalling the aid of these groups is an activity that has to be pursued with great sensitivity if the police are to maintain the credibility they require to act as effective promoters of public order.

It is useful to take the mass media as an example of the potential of multi-agency strategies. National and local television, radio and newspapers all have their own agenda when it comes to covering issues of social conflict, and their own reporting requirements and practices. These are frequently tempted to present issues in starkly polarised ways and can exacerbate the development of disorderly situations. Taking Smelser's model, we can see that the mass media may contribute to all stages in the development of conflict by:

a) increasing social strain, for example by presenting the issues involved in emotive terms;

b) playing up the severity of a situation;

c) assisting in the assembly of crowds by giving news of the venue;

d) helping communication between disorderly crowd leaders and possible supporters by reporting inflammatory rhetoric;

e) increasing demonstrators' sense of the universality of their emotion and their invincibility.

When attempting to liaise in a 'multi-agency' forum with other organisations and groups, the police run the risk of compromising the credibility of these in the eyes of their own members and supporters. Such a situation may well then damage the capacity of other potential sources of positive control to exert their own, independent influence on crowds. Openness and genuine negotiation are the keys to successful multi-agency liaison and there is critical discussion of this in Kinsey, Lea and Young (1986), Gordon (1987) and, in terms of the prevention of crime in general, Pearson *et al.* (1992).

The police as crowd and rioters

One point which should be remembered is that these theories can apply to all crowds including the police themselves. It is possible that a group of police officers may go through stages similar to those experienced by the participants in a disorderly crowd and lose control and react in an undisciplined manner. Feelings of anxiety and excitement that may generate rumour and hostile attitudes are understandable for police officers when faced with a 'mob', or even a crowd that they are ill-equipped to understand. When this happens they may over-react and engage in behaviour contrary to the spirit of legitimate civilian policing and in so doing themselves constitute a 'mob'. This is starkly illustrated in Lewis's refutation of Smelser's model in relation to the role of the National Guard at the Kent State University shootings in Ohio in 1970 (Lewis, 1975; D. Waddington *et al.*., 1989). Less dramatic parallels can be cited in England and Wales.

The key to avoiding such situations would appear to lie in police training. We would agree with Sadler in her study of the policing of demonstrations in Europe, that 'there is a fund of useful experience on which to draw to improve the chances of peaceful crowd control ..., and it is the responsibility of all parties concerned to build on the best experience of the past to avoid a more violent future' (Sadler, 1992:127-128), and hopefully too, to avoid resorting to the paramilitary option that is so often the norm in European demonstrations.

Further consideration of Smelser's model

Whilst we have used Smelser's work as a standpoint from which to view the social psychological dynamics of disorderly crowds, it will be clear that we are not comfortable with some of the more psychologically mechanistic aspects and implications that remain, despite his broadening the scope of his study to include the social factors that contribute to disorderly crowd behaviour. Additionally, although acknowledging that Smelser's acceptance of structural roots to conflict are an improvement on earlier theories, we would agree with D. Waddington *et al.* that:

> Underlying roots of conflicts are taken to be less important than the characteristics of the situation in which potentially conflicting groups find themselves. This reverses our order of determination, where conflict at the structural level is a basic precondition to disorder (D. Waddington *et al.*, 1989:174).

We have commented too on the implications for policing practice, emphasising that there is certainly much to be gained from a greater understanding of the social factors that contribute to crowd disorder. However, as Dunning *et al.* (1987:24) suggest, whilst mitigating, by reference to social structural factors, the excesses of Le Bon's theory, Smelser's work, with its emphasis on the idea that crowds are social disorganising environments, still retains the idea that crowds contain individuals whose violent response to social strains is essentially 'mindless'. This assumption has been authoritatively countered in the exploration of the disorderly crowd in history by writers such as Hobsbawm (1959), Rudé (1964) and Tilly (1979), who stress the economic and structural contributions to collective violence, and the instrumental nature of much riotous protest. Hobsbawm, for example asserts:

> The classical mob did not merely riot as protest, but because it expected to achieve something by its riots. It assumed that the authorities would be sensitive to its movements, and probably also that they would make some kind of immediate concession. For the mob was not simply some sort of causal collection of people united for some *ad hoc* purpose, but in a recognised sense, a permanent entity, even though rarely permanently organised as such (Hobsbawm, 1971:111, cited in Dunning *et al.*, 1987:25).

However, such accounts do not explain fully the contemporary forms of collective violence found in inner-city and peripheral estates in England and Wales. True, these do contain instrumental elements, but the immediate solution to problems rarely seems to be the dominant objective of the disorderly activity, which appears to focus on the destruction of private and public property and conflict with the police. This behaviour would appear to constitute a 'raising of issues', rather than to be directed at resolving problems or obtaining concessions. Such expressive violence has economic and political roots which are complex, but important to trace. In this connection we should not lose sight of the idea that violent crowd behaviour is essentially rational, even if the reasoning is tangled and is manifested in an expressive rather than directly instrumental form. Significant attention is given to this issue by B. Campbell in her study of recent disorders on housing estates, which are described in the following chapter. Referring to the outbreaks of disorder in Cardiff, Oxford and Tyneside she comments:

> These extravagant events were an enigma. They made world-wide news and yet they seemed to be powered by no particular protest, no just cause, no fantasy of the future. However, even in their political emptiness they were telling something about what Britain had become; the message in the medium of riotous assemblies showed us how the authorities and the angry young men were communicating with each other. The riots were the young men's way of speaking to their world (B. Campbell, 1993:x).

Smelser also fails to assess the impact of agents of social control, including the police, on a potentially disorderly environment, except, as mentioned above, by asserting his faith that firm, yet 'fair', action will be effective. D. Waddington *et al.* express this criticism in the following terms:

> Smelser counterposes the crowd, seen as unpredictable in its dynamics and predisposed to irrational behaviour, with a rational and disciplined state force whose only function is to control the crowd. The role of any kind of communication other than the directive is denied. Smelser refers to but does not analyse incidents which might be called flashpoints. Otherwise, interaction appears only as control exercised by the authorities (D. Waddington *et al.*, 1989:174).

We would argue that what may be seen by crowd participants as insensitive direction, or excessive force, can only enhance the sense of social marginalisation, mobilise bystanders ('observers' and 'cheerleaders'), and add credibility to techniques for legitimising further violent disorder as a form of expressive resistance (Craig, 1992). Thus we suggest that, in a broadly democratic policing environment, it is important to accept the idea of exercising minimal force and opening communication in the interests of facilitating the restoration of order and the public peace.

Whilst we have criticised the value of the 'riot curve' as a tool for enhancing police understanding, it does retain some force in that it draws attention to the necessity of managing the 'post-riot' environment - an issue largely absent from the work of Smelser and others. It is important that the prospects for policing in the aftermath of disorder are considered at all stages of planning and action. The way in which the police deal with disorderly, or potentially disorderly, circumstances is likely to be remembered by the community and requires that attention be paid not only to the future policing of the locality, but also to the activities of interest groups with or without an identifiable geographical location or affiliation. As we shall see, it is evident that this lesson is being learned, notably in the 'post-Scarman' period; however, it remains a moot point to what extent current thinking and developments are keeping abreast of changes in the nature of disorder and dissent.

As regards localised urban disorder in England and Wales, some success has been claimed, based on the experience of and innovation resulting from the Brixton riots. Following the outbreaks of disorder in 1981 and 1985, policing strategy has been substantially reviewed and efforts made to develop more positive, less threatening, day-to-day contacts between police and public. There have been efforts to reduce the emphasis on crime as the main reason for police contact with residents since research carried out locally showed that the public's demand for police assistance is 75% related to non-crime services. In inner-city and peripheral estate areas where crime rates are highest, the proportion of time spent by the police on non-crime related issues is likely to be lowest, unless there is a deliberate policy to shift the balance (Marshall, 1992:88). The response that has been developed in Brixton has involved the strategy of 'sector policing', aimed at enhancing the quality of contacts between police and public through face-to-face experience and negotiation, as described by Marnoch (1992). However, we have noted when discussing the impact of social control agencies on crowds and their development, that this, like all community-based strategies, is subject to considerable critical debate. We have also seen some evidence of successful co-operation in the facilitation of peaceable protest in the industrial relations field since the miners' strike of 1984/5. However, the plethora of discontents and conflicts that are developing as western societies experience economic restructuring and a fragmentation of confidence and interest in long-standing political institutions makes the task of securing a manageable consensus on public order policing practices increasingly problematic. Environmental issues and the attraction of people to alternative lifestyles which may bring them into conflict with conventional, or established, interests may be seen as areas of particular concern where recent police efforts to 'maintain' order have met with little success.

Conclusion

In conclusion, we would argue that a theory of crowd behaviour must accept the dynamic potential of crowd development and the essential, if increasingly more complex, rationality of participants acting within social structural constraints. A full understanding of these issues is essential to the development of an effective public order policing strategy and its tactical implementation. In the next section we will examine a number of instances of public disorder that have occurred in England and Wales. These examples throw further light on many of the questions regarding the understanding of crowds, their nature and the intentions of their participants that we have outlined here.

CHAPTER II
Instances of Public Disorder: changing policing strategy and tactics

Introduction

Our intention in this chapter is to give some indication of the changing strategy and tactics of policing large-scale potentially riotous situations and the changing form of dissent, from the late 1960s to the present in England and Wales. There are a number of qualifications that need to be mentioned at the start. First, we do not examine policing in Northern Ireland or Scotland here. The former we consider to be a special case in terms of the socio-political situation, the deployment of weaponry technologies and also the police-military relationship, whilst the latter has a different legal structure and policing organisation from those of England and Wales. Secondly, our selection of instances from the 1960s is meant to be more of a 'snapshot' than a comprehensive overview, with the purpose of indicating tendencies or directions of change. Our third qualification is that the anti-Vietnam demonstrations with which we start our examination are chosen because they are generally taken to be the first post-war instances of large-scale disorder. They are also the events with which a number of our interviewees were first engaged. We could equally, however, have taken examples from the Campaign for Nuclear Disarmament marches, or the late 1950s Notting Hill and Nottingham race riots (Rowe, 1994), or even the 'Teddy Boy' disturbances as our starting point, and to this extent questioning too the dominant perspective among both police and academics of a so-called 'golden age' of consensual policing in the immediate post-war period. Finally, whilst we are concerned with a relatively short period of time, we need to remember that riots and demonstrations are nothing new in British history. Indeed, it could even be argued from reading Thompson (1968), Pearson (1983) or Vogler (1991) that periods of 'calm', since for example the eighteenth century, may be the exceptions.

We do not see it as our task here to assess the *extent* of one determining factor or another for changes in the interplay between the police and what we shall loosely term demonstrators, although we do accept that policing strategies and tactics and the form of protest, to some extent at least, interact in causing that change. At the same time it would be misleading, on this basis, to assert that policing strategy, tactics, training or even development of technology, have been or are simply *reactive* to forms of protest. Such apparent 'reaction' by the police has been to a considerable extent *proactive*, i.e. it has been a response to a predicted or anticipated change in the form of protest or its intensity. What we shall be concerned with here especially is the extent to which policing strategy and tactics can lead to a situation of either escalation or de-escalation of conflict.

In the instances that we draw on, it is generally the case that the level of intensity of the situation was immediately affected by the form of police action or even, for some 'inner-city' riot situations, the (mis-)perceived form of police action, given the wider context. This may be taken as the 'flashpoint' of the incident, that is to say the action which was the immediate cause of escalation. As Lord Scarman made clear in his Royal Commission Report on the Brixton disturbances of 1981 though, the flashpoint *per se* is fairly meaningless without regard to 'antecedent conditions'. Indeed, D. Waddington *et al.* (1989) have suggested that there are six different 'levels' of analysis required for an understanding of a flashpoint. These levels, relating to structure and action, may be illustrated as a series of ever-widening concentric circles, ranging from the macro, or wider contextual factors, to the micro, or the immediate interactive situation, and are as follows:

	Level of analysis	**Factors**
1.	Structural	Material inequalities Inferior life chances
2.	Political/Ideological	Marginality Legitimacy
3.	Cultural	Protesters/Groups/Crowd Police
4.	Contextual	History Anticipation Liaison
5.	Situational	Police/Crowd organisation Objectives Spatial location
6.	Interactional	Flashpoint Arbitration/Escalation

(Sources: D. Waddington *et al.,* 1989:158, D. Waddington, 1992:26-27)).

We shall draw on D. Waddington, among others, when analysing our chosen instances. However, because of our emphasis on the relationship between the policing form and escalation, some distinction needs to be made here between the 'flashpoint' giving rise to an inflammatory situation, and the way that later situation is policed.

The changing forms of policing strategy and tactics could be shown on an historical progression or linear basis for large-scale public order situations generally, or in terms of specific 'types' of gathering. We feel that it will be more useful for clarification purposes if we adopt the latter approach for this empirical section, and the former when examining the rationale behind such changes in Chapter III. Accordingly, we utilise a four-fold typology of public order situation, namely:

1. Political
2. Industrial
3. Festival
4. Urban (Inner City/Estate).

These are not mutually exclusive categories, nor are they ranked in any sort of hierarchy. They will be examined in this order simply because of the date of the first instance in each group. The type of public order situation is, of necessity, loosely defined. For example, a traditional miners' rally, taking place at a time of industrial unrest in that sector could fall into the industrial, festival or even the political category. Similarly, whilst it will be argued later that different policing strategies are initially used for the various types of assembly, if the situations escalate, the tactics employed will generally converge. It may even be, as we suggest later, that these classifications, although conforming with present-day policing responses, do not encompass the newer forms of dissent, especially by those advocating 'alternative' lifestyles, environmental concerns and animal rights. In this sense, we could even suggest that the policing strategy currently employed against the latter group is inappropriate, in so far as it is based on an inappropriate typology. One important type of situation that we have consciously excluded from this study is that of sporting events.[2] These, we feel, deserve a book in their own right, because of the many developments in sports crowd control techniques, stadia technologies and policing and stewarding methods in recent years.

Most of the instances with which we will be concerned have been well documented by academics, police, commissions of inquiry or unofficial inquiries, and we will be using their reports as source materials. For the more recent disturbances we will be relying primarily on contemporary newspaper reports.

2. Dunning *et al.* (1987:26-27), however, do include the area of 'sport' and similarly identify four categories of social life for the differentiation of collective disturbances, namely 'politics, industry, organised leisure/sport and the community'.

1. Political demonstrations

Grosvenor Square anti-Vietnam demonstrations - 1967/1968

Whilst there were a number of anti-Vietnam War demonstrations focusing on the American Embassy in London's Grosvenor Square during this period, we will concentrate on three which may be seen to mark a distinctive period in public order policing experience and policy in England and Wales.

The first large-scale demonstration organised by the Vietnam Solidarity Campaign (VSC) was on 22 October 1967, protesting against American involvement in the war. About 5,000 people attended a rally in Trafalgar Square, and then a large number of these marched to Grosvenor Square to demonstrate outside the American Embassy. A police cordon was formed at one end of the square to seal off the Embassy, but ineffectively; missiles were thrown by some demonstrators and the confrontation escalated.

Traditional police tactics during this period involved either the 'trudge and wedge' or the linking of arms to form various kinds of cordons. Northam described these as generally involving 'large numbers of officers shoving themselves right up to a crowd in order to contain or move it, or forming running human wedges to break it up' (1988:88). However, he pointed out that this form of crowd control was no picnic for either the police or demonstrators. His words are relevant here, but also later when we discuss tactical developments with reference to Geary's work concerning the policing of industrial disputes:

> The potential of these tactics for causing injuries on both sides should not be underestimated. Those inclined to romanticise past confrontations as if they were no more than boisterous and gentlemanly line-outs, should remind themselves of (this) ... (Northam, 1988:88).

The next demonstration organised by the VSC was on 17 March 1968. About 25,000 assembled in Trafalgar Square for a march to Grosvenor Square. The National Council for Civil Liberties (NCCL) report on the demonstration suggests that the police repeated the mistake made in October 1967, namely ineffectively cordoning off the square (1968:1). The demonstrators actually broke through the cordon into the square itself, where there was another cordon before the Embassy and a large number of mounted police officers became targets for missiles. Eventually, the mounted police were deployed to clear the square, resulting in many injuries to demonstrators and police in the ensuing panic and confusion. One NCCL official observer reported that:

> 1. The conduct of some officers was exemplary - to remain cheerful under a shower of earth sods and under constant pressure from a mob is no easy thing.

> 2. The conduct of most officers was tough but fair. The conduct of a few was angry and violent. Police should use minimum force and must not hit a demonstrator when he/she is trying to get away.
> 3. The strategy to control the demonstration was ill-conceived and ineffective. The use of police horses is to be regretted most strongly (NCCL, 1968:6).

There were 250 arrests, in what *The Economist* called 'Britain's worst political riot since the war' (cited in Fowler, 1979:57).

The third demonstration took place on 27 October 1968 and violence was widely anticipated. *The Times*, for example, said on 5 September:

> Police and leaders of the peaceful demonstrators share the view that 27th October could bring the most violent upheaval in Britain for many years. Some senior officials believe that they are faced with a situation potentially as violent as the student demonstrations in Paris and Berlin earlier this year (cited in Fowler, 1979:58).

However, in the event this demonstration was markedly different from the previous two. About 60,000 demonstrators assembled for a march from Charing Cross to Hyde Park for a rally. The march had been well publicised and the organisers had stressed that the march should be peaceful and avoid Grosvenor Square. There were 8,846 police officers present, out of a total Metropolitan Police complement at that time of 20,000 (Bowden, 1978b:211). The police displayed a more accommodating attitude, controlling or diverting traffic, flanking both sides of the marchers, and keeping a low profile for reinforcements and mounted police. Fowler argues that the police in this instance also 'turned a blind eye' to minor skirmishes and missiles (1979:62-63).

The main march was generally characterised by 'non-violence and cordial relations' between police and demonstrators (D. Waddington *et al.*, 1989:57), although two groups broke away at one stage and stormed towards Grosvenor Square. However, this time the police cordon was not breached, nor were police truncheons or horses used. The main march concluded in the evening with both police and demonstrators linking arms and singing 'Auld Lang Syne'. James Callaghan (at that time Home Secretary and later to be Prime Minister) called it 'a demonstration of British good sense' (Fowler, 1979:60), whereas Fowler went on to argue that the:

> ... rapport between police and protesting public ... (and indeed) the British acceptance of their police ... helps the policemen themselves to do without para-military paraphernalia of crowd control like the water canon and the tear gas canister (1979:63).

D. Waddington *et al.*, on the other hand, argue that these three demonstrations represent 'turning-points' in police-demonstrator 'consensual' relations. First, they represented an escalation in the form of dissent, from non-violent action and passive resistance to 'unprovoked attacks on the police' by a minority of demonstrators (1989:56). However, despite such 'direct violence', there was an 'apparent success of the tactics employed' by the police at the third demonstration. For the police themselves though, D. Waddington *et al.* argue that the perception was rather 'that they could no longer rely on traditional methods of crowd control' (1989:57). This led to the development of new tactics, such as snatch squads, and the creation of the **Special Patrol Group** in the Metropolitan region.

National Front march: Lewisham - 1977

On the 13 August 1977 the National Front (a neo-fascist organisation) had planned a march from New Cross to Lewisham (in south-east London), an area with a high proportion of African-Caribbean residents. There were pleas by the local council, the Trades Union Congress, the Labour Party and even sections of the media for the march to be either re-routed or banned. Otherwise the likelihood of confrontation between the marchers and black youth and anti-fascist groups was high. However, the march was allowed to take place as planned.

The number of National Front demonstrators was relatively small and these were periodically attacked by those counter-demonstrators who managed to break through the police cordons. Eventually the police had to re-route and then stop the march. Rollo states that:

> it was only after the demonstration was over that the police attack was launched, with riot shields and SPG vans driving at top speeds towards groups of demonstrators, and repeated charges on foot and horseback. The violence of those clashes caused a national furore (1980:185).

Certainly, this was the first time that shields had been deployed on the British mainland, and the mere fact that they were there in readiness prompted questions as to the extent to which the 'police explicitly decided upon a confrontation strategy'. Jefferson suggests that 'if the Notting Hill carnival of 1976 (to which we refer later) promoted the acquisition of special protective equipment, Lewisham highlighted the issue of special protective clothing' (1990:6). In respect of future strategy and tactics, Lewisham can reasonably be seen as a 'watershed'. The Commissioner of the Metropolitan Police, McNee, stated some time later that:

> the shortcomings of the traditional helmet were evident during the (Lewisham and Notting Hill) disturbances (a month later) ... and I stress that it does not mean we have forsaken traditional methods of policing demonstrations (cited in Rollo, 1980:186).

Poll Tax demonstration: Trafalgar Square - 1990

On 31 March 1990, a day before the introduction of the Conservative government's reform of local government finance from a system based on rates (paid by all adult householders according to the value of their property) to the community charge (a flat-rate 'poll tax' for all adults regardless of the value of property but calculated according to local council expenditure), the All Britain Anti-Poll Tax Federation (ABAPTF) organised a protest march in London along Whitehall (passing Downing Street) to a rally in Trafalgar Square. This had been preceded by a number of meetings, protests and clashes between police and demonstrators nation-wide, for example in Bradford, Birmingham and Bristol and, on 9 March 1990, in Lambeth, London where there was sporadic disorder after a meeting of about 2,000 people outside the town hall.

This background was obviously taken into account in police contingency planning for 31 March. According to the *Today* newspaper of 2 April (cited in D. Waddington, 1992:23), Special Branch police officers had learned of a proposed attempt by the Socialist Workers Party to mount some kind of concerted attack on 10 Downing Street (the official residence of the Prime Minister) in connection with the march. The demonstration organisers themselves had, however, been in negotiation with the police before the event concerning the route of the march, the number of stewards and the estimates of the possible number of demonstrators. The latter varied between 20,000 and 58,000 (D. Waddington, 1992:23).

The majority of marchers passed peacefully through Whitehall to Trafalgar Square, although 200-400 occupied an area opposite Downing Street as a 'sit-down protest'. According to the organisers, it had been arranged with the police that if this were to happen, the rest of the marchers should be encouraged to proceed to Trafalgar Square. In the event it would seem that the majority did continue moving, but after some clashes between the sit-down protesters and the police, the police mounted a horse-charge. The words of one demonstrator clearly convey the frustration of the 'trapped' demonstrators: 'We had nowhere we could go. There were barricades stopping people going (any) further in Whitehall. There were lines of police stopping people going anywhere else and then there were police horses' (D. Waddington, 1992:24, citing a Channel 4 TV production 'The Battle of Trafalgar' - broadcast 18 September 1990). Demonstrators were dispersed both south and north, the latter into Trafalgar Square, where at that time 'a carnival atmosphere had so far prevailed' (D. Waddington, 1992:8). Disorder then broke out there too, either on the arrival of two police vans or of police in riot gear. Police were then:

> bombarded with missiles procured from a building site, a workman's hut was set on fire and South Africa House was stoned and almost set ablaze. Three police vans sped through the crowd at 30 to 40 m.p.h., causing demonstrators to flee for their lives. Mounted police tried to clear the square, driving some demonstrators along Charing Cross Road and St. Martin's Lane. However, as the crowd moved off, some of its members looted the windows of West End shops (D. Waddington, 1992:8).

Policing strategy and tactics in this instance stand in marked contrast to those of 1967/8, as does the form and intensity of the dissent being expressed. By this period, there was a more developed possibility for contingency planning by the police, with the **Public Order Act (1986)** requiring formal advance notification of planned demonstrations, route, etc., a more sophisticated Command and Control system (incorporating the '**Gold, Silver and Bronze**' decision-making structure, to which we refer later), as well as new technologies such as evidence-gathering teams for prosecutions after the event, and riot tactics involving the deployment of horses, shields, snatch squads, **Police Support Units (PSUs)** and police land rovers. Despite this, however, the Metropolitan Police Debriefing Report admits that some of the blame for the escalation must be placed on shortcomings in the co-ordination of tactics and in communication among the police themselves and, in terms of auditory equipment, in communication between them and the demonstrators (1991:13-15). The outcome of this has led to a reassessment of Command Band training, which again is something that we refer to later.

2. Industrial disputes

Introduction

The instances of industrial disputes that we have chosen illustrate well the changing relationship between police and pickets from the early 1970s to the late 1980s and also the forms of action they employ. The example of Saltley, during the miners' strike in 1972, also helps to put the policing of Orgreave in 1984 into context.

Geary has argued that the policing of industrial disputes in Britain, up until the 1984/5 miners' strike, displayed 'historical trends' towards non-violence on the part of both the police and the demonstrators. Indeed, he argues that by the period with which we start our examination of these changes:

> industrial confrontation became a sophisticated political game in which two sides, police and pickets, battled for public opinion. This meant that violence beyond mere pushing and shoving was generally regarded as counter productive in strike situations (1985:133).

Geary identifies, from the late nineteenth century up until 1984/5, six stages or 'patterns' of confrontation involving distinct forms of both control tactics and disorderly conduct, each of which follows an increasingly pacific trend, moving from shooting as the dominant control tactic to stoning as the dominant form of disorder, through batoning and violent/non-violent picketing and mutual pushing and shoving to symbolic confrontation. The reasons for the reduction in violent confrontation are, he suggests, 'the constitutionalisation of strikers and police, the growth of the mass media and the democratisation of civil liberties' (1985:146) plus changes in the method of policing public order. Geary suggests that the 1984/5 form may simply be a 'one off'. As for the future, there is every reason to suppose that the violence of the miners' strike will not generally be repeated in subsequent disputes'; however, he warns that the 'constraints on violence have, to some extent, been offset by modern police tactics which tend to generate a vicious spiral of violence and destruction' (1985:147).

Jefferson (1990) has criticised Geary on two main grounds: first, for a 'romanticised' notion of 'history as progress' instead of seeing confrontation as being contingent on the extent/form of consensus and hegemonic crises (Jefferson also incorporates the problem of 'technological drift' as a determining factor), and second (as mentioned earlier) for having an over-idealised notion of 'push and shove' as a consensually benign public order policing tactic.

Saltley - 1972

Following a breakdown in the wage-bargaining process and a membership ballot, the National Union of Mineworkers (NUM) called a strike from 9 January 1972. This was the first national coal strike since 1926. Because of the length of the prior negotiations, coal reserves held by both the National Coal Board (NCB) and the Central Electricity Generating Board (CEGB) were considerable. Accordingly, picketing was directed at preventing not only the production of coal, but also its further distribution from NCB and coke depots to power stations.

One month into the strike, Saltley Coke Depot in Birmingham was, according to Geary 'the only major distribution point still operating'. A token picket there of 12 men had succeeded in turning back only about 100 distribution lorries of the estimated 800 per day (1985:73).

There were a number of disorders occurring during this period at various picketed sites, mainly resulting from excessive 'push and shove'. On 3 February, however, a picketing miner from Yorkshire, Fred Matthews, was killed by a non-union lorry driven at speed through a picket line near Scunthorpe. The next day mass picketing began at Saltley, with about 200 pickets. On 5 February strong feelings were aroused, given the Matthews tragedy, when a lorry was driven through both pickets and police, but Gas Board officials responded to the mood and de-escalated the situation by closing the depot gates until the following day. On 6 February, however, about 2,000 pickets assembled facing 400 police officers. On 10 February these numbers had increased to the clearly escalatory level of about 15,000 pickets (from different mining areas in Britain) and 800 police. Given the difference in numbers, the Chief Constable of Birmingham, Sir Derrick Capper, decided to close the gates of Saltley 'in the interests of public safety' (Geary, 1985:76).

This was seen by the miners as a resounding success of their 'flying picket' strategy and it meant that the strike was nearly over. Because of alleged coal shortages, the

Conservative Heath government declared a state of emergency on 9 February, followed by a reduction in the working week to three days in order to save power; and the strike ended on 25 February. Geary argues that 'little more than spirited pushing and shoving seems to have taken place' where the occasional scuffles between police and pickets occurred (1985:77). However, many sources suggest that Saltley went into the annals of the police and the Conservative Party as a defeat. This is well illustrated by the Welsh Campaign for Civil and Political Liberties (WCCPL) who refer to a letter written subsequently by Reginald Maudling, then Home Secretary:

> I remember when during the miners' strike pickets threatened to close the Birmingham Coal Depot, and, in fact, succeeded in doing so, the then Chief Constable of Birmingham assured me that only over his dead body would they so succeed. I felt constrained to ring him the next day to enquire after his health (WCCPL, 1985:86).

We shall refer to Saltley again later, as an event that served to condition the manner in which the 1984/5 miners' strike was perceived politically and in terms of policing strategies.

Certainly, important moves towards changing police strategies and tactics for dealing with potential public disorders were started from this point. Mutual Aid between police forces nationally, which had only been established bilaterally with the Police Act (1964) (McCabe and Wallington, 1988:45), was developed and strengthened, a revamped Civil Contingencies Unit of the Cabinet Office was established, a review of public order training was undertaken and **PSUs** for each constabulary and the National Reporting Centre (NRC) (now **Mutual Aid Co-ordination Centre (MACC)**) at New Scotland Yard were created. Also, new legislation was introduced to curb 'flying pickets'.[3]

The Grunwick dispute - 1976/1978

Still within Geary's mutual 'push and shove' stage is the Grunwick dispute. However, there have been a number of suggestions that Geary seriously underestimated the level of violent confrontation between the pickets and the police, especially by putting Grunwick on a par with Saltley in 1972. Grunwick is especially significant in that it involved the first mass picket since that deployed during the dock strike of 1972 following Saltley.

The strike at the Grunwick (photographic) Processing Laboratories in Brentford, north-west London, started on 23 August 1976 and lasted until July 1978. The initial dispute was about the alleged unfair dismissal of one of the workers but it then became a contest over the right to join a trade union. Striking staff, slightly under one-third of the workforce, approached the Association of Professional, Executive, Clerical and Computer Staff (APEX) which then applied to the Advisory, Conciliation and Arbitration

3. Secondary picketing was made unlawful, i.e. under civil not criminal law, by the Employment Act (1980), but police may intervene and limit numbers where they fear serious public disorder under the **Public Order Act (1986)**.

Service (ACAS); and ACAS opened discussions on the recognition of the union by the Grunwick management but to no avail. On 2 September, after APEX had declared the strike 'official', Grunwick dismissed all the strikers for breach of contract.

Eventually a mass picket was called for to start on 13 June 1977. On the first day this consisted of about 700 people who obstructed the factory entrance simply by weight of numbers in the narrow streets. When they failed to keep to the pavements as requested, the police 'cleared the road and formed a cordon' using, it was alleged, 'very aggressive tactics' (Geary, 1985:85). These were apparently repeated each time an obstruction occurred, resulting in many 'scuffles and arrests'. Thereafter, employees still at work were delivered through the gates by bus and the police employed 'snatch squads' for arrests. The 'push and shove' intensified at the sight of the buses. Subsequently the level of violent confrontation varied according to events. For example, in March 1978 the Grunwick management failed in its attempt to gain a High Court injunction against the picketing of retail outlets using its services. In May convictions against six strikers were quashed and they were awarded £3,500 damages against the police (Scraton, 1985:153).

On 21 June a number of Members of Parliament from the Labour Party (then the party in government) attended to support the picket and one, Audrey Wise, was arrested for obstruction. Another MP, Jo Richardson, said that she 'had never seen such open intimidation and brutality from the police as at Grunwick' (Rollo, 1980:183). Arthur Scargill, leader of the Yorkshire miners, was arrested for obstruction on 23 June, but later acquitted. On that day too a police officer, PC Trevor Wilson, was severely injured when hit on the head by a milk bottle. After this incident, Lord Scarman (whom we will refer to later in connection with the 1981 Brixton disturbances) was appointed to lead an inquiry into the whole situation.

The confrontation and the numbers of pickets seemed then to decline; but there were two further serious events. The first of these was on 11 July, a national day of action, when about 18,000 pickets and 3,500 police officers attended (Clutterbuck, 1978:191), and the second was when the last mass picket was to be held, on 7 November. This latter, Geary argues, involved 'some of the most violent clashes of the whole dispute' (1985:88). Later, a deputation consisting of about 3,000 people marched to the local police station to protest at the police precipitation of the violence.

Clearly, there are similarities between the forms of protest and policing of Saltley and Grunwick. However, the geographical space differed, as did the composition of pickets (those at Grunwick included a large proportion of students, for example) and the organisation of the policing. At Grunwick, for example, the Metropolitan Police Special Patrol Group were continuously deployed during the mass picket between 13 June and 11 July 1977 and, it would seem, had an offensive role which brought a wave of protests; even the Trades Union Congress called for a Home Office inquiry into its conduct.

Clutterbuck (1978:200-201) gives the police estimates of the maximum number of both police officers and pickets who were present during the mass picket. This shows what would now be regarded as an enormous police presence in relation to the number of pickets, ranging from 1:4 to near parity:

	Police	Pickets
13 June 1977	308	700
14	508	700
15	356	700
16	521	700
17	688	1,500
20	556	1,200
21	765	1,000
22	631	800
23	793	2,000
24	1,521	2,200

Orgreave - May 1984

The miners' strike 1984/5 started on 12 March primarily over the issue of pit closures and reductions in the workforce, but also linked to the definition of what constituted an uneconomic pit. For four months prior to this there had been a national overtime ban. But matters were brought to a head with the announcement by the National Coal Board (NCB) on 6 March that its planned budget for 1984/5 entailed the closure of 20 more pits (the same number as had been closed during the preceding year) and the loss of 20,000 jobs. The National Union of Mineworkers (NUM) insisted that there were 'secret plans' to close down an additional 50 pits, making a total job loss for the industry of 75,000 (McCabe and Wallington, 1988:18).[4]

The Yorkshire miners were the first to strike and were followed by miners in South Wales, Scotland and England (Kent and Durham). The majority of mines in Nottinghamshire continued working so that most of the picketing which took place was concentrated in the Nottinghamshire area.

The policing of the strike involved a number of novel features in terms of organisation, decision-making and the use of what have generally been regarded in hindsight as escalatory tactics. The latter especially will be highlighted in our later discussion of policing at Orgreave. New policing tactics also included the setting-up of road-blocks, which were used occasionally in serious criminal investigations, but for which there was no precedent in situations of industrial unrest. A number of constabularies instituted

4. In the event, this number of feared job losses became reality, as McCabe and Wallington show: 'Within two years of the end of the strike, 60 of the 170 collieries open at the start of the dispute had closed, and 78,000 of the work-force had left the industry' (1988:108).

road-blocks against potential pickets following, first, a statement in the House of Commons by the Home Secretary that:

> The mere presence of large numbers of pickets can be intimidating. The police have a duty to prevent obstruction and intimidation and enable those who wish to go to work to do so. They have the power to disperse large numbers of pickets and to take preventive action by stopping vehicles and people (McCabe and Wallington, 1988:59).

Secondly, an announcement was made in the House of Commons the following day by the Attorney General that 'the police are entitled to stop pickets if they fear a breach of the peace, (WCCPL, 1985:49); and, thirdly, Kent Constabulary set up road blocks against potential picketing miners attempting to travel north through the Dartford Tunnel.

As to new policing organisation, the miners' strike saw the first major use of the **National Reporting Centre** (set up after Saltley), especially for large-scale forward planning, direction and co-ordination of mutual aid arrangements. On 19 March, David Hall, Chief Constable of Humberside and controller of the NRC, referred to some 8,000 police officers having been mobilised, which constituted 'the most mobile and sophisticated police reserve yet assembled in Britain' (Geary, 1985:137).

Despite this build-up, by 23 March there had only been 96 arrests, and these mainly for obstruction, although Hall warned at this stage of the possibility of the police being issued with riot shields (Geary, 1985:137). By 11 December, however, 7,314 people had been charged with offences relating to the strike (WCCPL, 1985:107).

Before we examine the events at Orgreave, it needs to be said that a number of authors suggest that these should be considered in the context of Saltley 1972, mentioned earlier. Certainly, the Assistant Chief Constable of South Yorkshire argued that 'there is no chance that this (Orgreave) will be the Saltley of 1984. The plant will remain open until British Steel decides otherwise' (D. Waddington *et al.*, 1989:84). Also the President of the NUM, Arthur Scargill, when calling for a mass picket of Orgreave referred to 'doing a Saltley' (D. Waddington *et al.*, 1989:84). However, whether Saltley was in the minds of the majority of police strategists or even pickets is debatable.

Towards the end of May 1984 it became generally known that the British Steel Corporation (BSC) was intending to move a large amount of the coke stored at its coking plant at Orgreave, South Yorkshire. On 23 May there were approximately 25 pickets in attendance but by 18 June there were about 7,000. The first 'flashpoint' occurred on 27 May, when Scargill was apparently knocked over by police and later called for a mass picket. On 29 May there were approximately 1,500 pickets and 1,700 police officers from 13 different forces. Mounted police baton-charged pickets, for the first time since the 1940s, three times during the day and **PSUs** in full riot gear (including long shields and NATO-style helmets) were deployed, as were police dogs.

There is some dispute at to when the mounted police and PSUs were engaged. According to a report by the Chief Constable of South Yorkshire, Peter Wright, to his local Police Authority, this was 'only after officers in normal uniform had been subjected to a prolonged barrage of missiles' (McCabe and Wallington, 1988:76). Other sources suggest that only minor disorder had occurred before the first charge by mounted police officers. Geary suggests that missiles directed by the pickets against the police included 'smoke bombs, firecrackers, fence posts, stones and bricks' (1985:139).

On 30 May the conflict was even more intense, especially following the arrest of Scargill for 'obstructing the police and the highway'. But, on subsequent days the strikers reduced their presence to a token picket. However, on 18 June, the hundredth day of the strike, a 'secretly organised mass picket' of Orgreave took place (D. Waddington *et al.*, 1989:87) which was clearly anticipated by the police. Some sources even suggest that it was a case of planned confrontation on the part of the police in retribution for Saltley. For example, rather than pickets being subject to road-blocks, 'not only had Orgreave been left open, but alternatives had been closed off' (striking miner quoted in D. Waddington *et al.*, 1989:89). In any event, 7,000 pickets and 3,400 police officers (including 181 PSUs, 50 police horses and 58 police dogs (Northam, 1988:55)) 'battled' at Orgreave that day. Short shields were used for the first time during the strike, and 93 pickets were arrested and charged with riot and unlawful assembly.[5] After this confrontation, the miners' strike gradually collapsed through a slow return to work, but this also marked the beginning of heightened conflict between striking miners and police, and between striking and working miners in their communities.

Northam refers to when, in 1985, the BBC Television documentary programme *Brass Tacks* invited John Alderson, former Chief Constable of Devon and Cornwall Constabulary, to view and comment on the official police video of events at Orgreave on 18 June 1984. We quote Alderson at length here as his comments are especially important on the problem of the relationship between policing tactics and escalation, and also because he brings into the equation the doctrine of 'minimal force', which we discuss in the Introduction:

> *Brass Tacks*: You have looked at the police video of Orgreave. Judging from it, who attacked first?
>
> *Alderson*: I think it is fair to say that although there was pushing and shoving by the miners and one or two throwing missiles of one kind or another, the general escalation, the first escalation it seems to me, came from the cantering of police horses into the crowd which merely heightened the tension and increased the violence. Which is contrary to what the police stand for. The

5. All of the charges were eventually dropped by the prosecution and the accused acquitted on the 48th day of the trial.

police are there to diminish the violence, not to increase it ... Look what has happened. The commander on the spot has exercised one of his options, and that is his option to release the mounted police to charge in and to intimidate the crowd, in order to drive them back and relieve the pressure.

Brass Tacks: Would you have done that?

Alderson: I personally would not have done that at this stage because what is happening now is that this is causing the crowd, which is already tense and angry for reasons apart from on this particular day, now to become even more angry and respond you see. Now, you will find that after that mounted police charge, the throwing of stones at the police increases a little, though not, at this stage, on any great scale. So that has merely provoked anger and reaction and escalated the day's proceedings higher than I would have wanted to do at that stage.

Brass Tacks: How would you describe the decision to send in the police horses there?

Alderson: I would describe it as the sort of thing that you might read in a manual, but on the spot this is where judgement comes in. If you are trying to police with minimum force and get away from this field today with few casualties, then the police should not start the escalation on any scale (Northam, 1988:56-57).

Alderson was also one of the contributors to the McCabe and Wallington Report, referred to earlier, which argued that:

> Evidence as to the policing of the miners' strike during 1984 and 1985, supplemented by subsequent developments, indicates mistaken emphases in the approach to policing situations of potential disorder, especially industrial disputes, embracing training, assessment of priorities, and policing strategies (1988:3).

One of the 'subsequent developments' was the Wapping dispute, to which we now turn.

The Wapping dispute - 1986/1987

The Wapping dispute involved printworkers in two unions, the Society of Graphical and Allied Trades (SOGAT) and the National Geographical Association (NGA), initially protesting against the News International (NI) Group's planned introduction of new technology and, later, seeking reinstatement of dismissed workers. The unions and NI had been in negotiations for some years concerning the introduction of new printing technology without reaching agreement. According to McCabe and Wallington, not only did this give NI ample time to prepare for a battle, but it actually took steps to precipitate direct confrontation. First, it built a plant at Wapping in London (supposedly to house a new evening newspaper but, it later transpired, transfer of the whole news operation). The new building, which Mason describes as a 'fortress' (1986:2448), was surrounded with razor-wire and had utilised sophisticated gate and internal security arrangements and floodlighting. Secondly, elaborate preparations were made for Electrical Electronic and Plumbing and Telecommunications Trade Union (EEPTU) members to operate the new print technology; for National Union of Journalists members to continue working at the new plant on the basis of special 'financial incentives'; and for the distribution of the newspapers to be transferred to non-print trade union workers at Thorns National Transport Roadfreight UK (TNT) and away from British Rail, whose staff may possibly have engaged in sympathetic 'blacking' (McCabe and Wallington, 1988:111).

NI insisted on a number of conditions for print union involvement in the Wapping plant, including that they end their restrictive 'closed shop' practices, that they agree to flexible rostering and that all employees sign five-year no-strike agreements. A strike was called by the unions to start on 24 January 1986, and the 5,000 union members obeying the call received instant dismissal from their jobs. All NI operations were then transferred to Wapping, where twice-weekly mass picketing started. Police set up road-blocks, a number of which remained permanently throughout the dispute, in order to assist the 'free movement' of TNT lorries.

The first major problem occurred on 15 February. After a march of women printworkers, about 3,000 demonstrators assembled outside the plant where there was a police presence of 1,029 officers, including eight SPG units. In the conflict which ensued, horses and riot equipment were used for the first time (London Strategic Policy Unit (LSPU), 1987:9). By May, feelings of frustration were running high. Not only was the picketing having no effect, but up until then the strikers had been receiving strike pay from SOGAT funds which ceased because of court sequestration. On 3 May there was a 'large and angry demonstration' involving approximately 10,000 demonstrators and 1,744 police officers. After about 12 missiles and a smoke bomb had been thrown at the police cordon 'the police responded with a resort to riot-control tactics' (McCabe and Wallington, 1988:112). Police were severely criticised by Members of Parliament and trade union leaders, who had been in attendance for the rally, for over-reaction on such a scale.

Following this there were a number of violent confrontations and also peaceful meetings. On 31 July a Court injunction was obtained by NI limiting the number of pickets permitted at the plant to six, but violent confrontation still occurred periodically. On 24 January 1987, however, the first anniversary of the dispute, there was a mass rally at the plant of 15-20,000 demonstrators and about 1,000 police officers. What followed were the worst scenes of violence yet. After some had started throwing stones at the police and a lorry had been overturned:

> the police response was a full-scale sweep of mounted and riot-equipped officers moving towards the platform and the main demonstration. In the panic which followed many were caught in the line of advancing police horses or fell victim to the stampede away from the riot police. Over 300 injuries were recorded, including 167 police officers, and there were 67 arrests (McCabe and Wallington, 1988:113).

The example of Wapping has been chosen to illustrate the dramatic escalation of violent confrontation which has occurred in industrial disputes. That escalation has, of course, been reflected in increasing recourse to the courts and legal proscription. It is important to stress here, however, that although policing strategy and tactics relating to industrial disputes have obviously undergone considerable change, this does not mean that there is in such disputes inevitably a situation of immediate confrontation, rather than conciliation and accommodation between police and pickets/demonstrators. Indeed, a number of strikes since 1987, and more recent demonstrations by miners in London, have been peaceful and relations have been cordial. It is unlikely, however, that either the miners' dispute 1984/5 or Wapping 1986/7 can simply be viewed as 'isolated instances' against the trend.

Another problem arising from Wapping, which needs to be taken into consideration, is the toll it took on police-community relations both locally and within the Metropolitan Police area generally. The method of policing the dispute, involving a large number of permanent and temporary road-blocks to allow TNT vehicles 'unhindered access', for example, had the contrary effect for local residents. This lost the police a great deal of local support and caused a number of demonstrations by the residents against both the restrictions and the plant. The toll it had on preventive and service policing generally is indicated by the Metropolitan Police Commissioner Sir Kenneth Newman's report to the Home Secretary in January 1987 in which he said that: 'The deployment of an average of 300 officers a day has eroded the number of officers available for divisional street duties, thereby preventing them from making a contribution to other aspects of the force goal' (LSPU, 1987:33).

3. Festival disorder

Introduction

We suggested earlier that a major watershed in the development of the contemporary method of policing, at least regarding the use of the shield as defensive weaponry, was the Notting Hill Carnival in west London of 1976. The policing of the riotous situation that developed there is commonly and graphically described as the 'dustbin-lid and milk-crate' stage of riot technology. The policing of the 1976 carnival and that of the following year provide contrasting examples of developments in the policing of festivals: one employing high-profile 'saturation' policing, but no defensive weaponry other than the traditional, relatively short, baton, the other (initially at least) using a low profile approach and relying heavily on prior negotiation and self-stewarding, but with reserves on stand-by with reinforced helmets and shields.

While the following instances have been chosen to illustrate watersheds in the development of the strategy and tactics of control, it should be noted that celebrations of differences of ethnic orientation, culture and lifestyle are an increasingly significant aspect of contemporary society. Further, the successful aspects of the development of strategy and tactics employed in Notting Hill may, with due appreciation of the diversity of forms such celebrations may take, be read as indicative of a potential, at least, for the development of a positive approach to a range of other groups who are presently marginalised, or seen to be in conflict with established interests. This is an issue that we will address more fully in the Conclusion.

Notting Hill carnival - August 1976

The Notting Hill carnival had by 1976 been held annually for 10 years over the August Bank Holiday weekend. It is organised primarily by African-Caribbean residents of Notting Hill Gate, an inner-city area of west London, but usually attracts crowds of about 250,000 of all races over the period of the festivities. Until 1976 the carnival had remained largely peaceful. However, in that year, there was a determined policy on the part of the Metropolitan Police to crack down on pickpocketing at the event by targeting black youths. The policy entailed saturating the area with uniformed officers, initially numbering 903 and on the following day 1,598 (Borrell, 1976). David Leigh, correspondent for *The Times* newspaper, reported that one officer, Chief Superintendent Ron Paterson, had informed him, only two hours before the first riot started, that: 'We were not here in sufficient force last year. This time we have enough men to deploy where we want' (Leigh, 1976b). Similarly, after the event, Sir Robert Mark, Metropolitan Police Commissioner, when defending the decision to deploy such numbers of officers, and also the move to contain the riot by cordoning off the whole area, argued: 'There are not going to be any 'no-go' areas in the Metropolitan police area and there is no question of us abdicating our responsibility' (Borrell, 1976). On the preceding day, there had been serious assaults on three police officers, including one requiring 30 stitches for a 'razor wound' and one being 'knifed in the groin' (Borrell, 1976). Leigh reported that he saw 'lines of police officers' such as he had 'only seen otherwise during National Front marches' (1976b), and quoted a community relations officer as saying that: 'Until this weekend there has never been so many police around on duty ... I'm sure this caused some of the young blacks to over-react ... Until this year, police maintained a low profile' (Leigh, 1976a).

It is unclear what 'flashpoint' changed the situation from sporadic scuffling to riot. Kettle and Hodges suggest this was when a white woman was physically attacked and her handbag stolen (1982:79), whilst others suggest that it was due to the wrongful arrest of a pickpocket (Tendler and Leigh, 1976; Joshua and Wallace, 1983:61). Be that as it may, as Leigh describes developments, the 'police met a hail of bottles (and later bricks) ... using milk crates and dustbin lids to protect themselves' (Leigh, 1976b). There were eventually 60 people arrested, 50 of whom were black. The riot, involving, Mark suggested, not more than 800 'hooligans', resulted in 325 police and 131 civilians being reported injured; 35 police vehicles and 4 ambulances being damaged; 2 British Transport Police cars being burnt out; and damage being caused to 321 premises with 3 shops looted (Borrell, 1976).

Notting Hill carnival - August 1977

The following year, the arrangements for policing the carnival were significantly different. First, the police had been in consultation with festival organisers who instituted a system of self-stewarding. During the day-time events, it was stewards who made most of the incursions into the crowd to 'rescue white people and drag out the trouble-makers' (Tendler and Huckerby, 1977). Secondly, there was a policy, again during the day-time, of low-profile policing. David McNee, the new Commissioner of the Metropolitan police, specifically stated that his officers 'would remain unobtrusive unless trouble broke out' (Tendler and Huckerby, 1977). However, each evening an increasing number of officers appeared with new 'strengthened' helmets and by the second evening riot shields (these having been first deployed at Lewisham in South London earlier that month and then at Ladywood in Birmingham at National Front demonstrations). By the end of the carnival rioting occurred again, resulting in a number of injuries, looting and fires (Tendler and Huckerby, 1977; *The Times*, 1977).

The aftermath

Despite manifest change in policing strategy and tactics after 1976, the carnival failed for some years to return to its pre-1976 state of well-being. An increasing number of police officers were deployed and on 'stand-by'. Kettle and Hodges suggest that by 1979 and 1980 these amounted to almost 50% of the Metropolitan Police; 1979:10,135; 1980:11,022 compared with a total Metropolitan Police complement in 1980 of 23,691 (1982:80).

A decided reorientation in policy, however, seems to have occurred in 1987 with the appointment of Larry Roach as Commander (Operations) for west London with special responsibility for the carnival. He suggests that a two-pronged strategy was undertaken; first in the community itself through an 'intense multi-agency commitment' to the prevention of criminality (in the form of hard drugs especially) and disorder (Roach, 1992:105); and, secondly, in terms of the carnival *per se*, rapid response PSU contingency planning.

In 1987, there was in fact serious rioting and 1,161 reported crimes, including nearly 400 'thefts from the person or robbery', and the deaths of two people, one 'at the hands of a robbery gang' (Roach, 1992:102). However, Roach argues that by 1988 his policy of community involvement and preventive policing, based on 'professionalism and restraint' and also the deployment of 'sensitive' 'arrest squads of specially trained officers' had taken effect. In 1988 there were only 193 reported crimes (Roach, 1992:105-106), and since 1989 there have been no serious incidents of disorder (P. Waddington, 1994:19).

4. Urban disorder

Introduction

Of all the types of public disorder that we are examining here, urban disorder is the most worrying for the population generally because of its apparently *spontaneous* nature. It is this which is the main incentive behind the drive on the part of the police to develop effective community 'tension indicators', referred to later. It can be argued that whilst such disorder is largely precipitated by police action, it generally reflects a wider breakdown in social consensus in the area in question, which may be due to a variety of factors.

The areas with which we are especially concerned here are loosely classified as '**inner-city**' and '**estate**'.[6] Both are characterised by a history of social deprivation, large-scale unemployment, and political marginalisation. The inner-cities have significant African-Caribbean or South Asian populations, whereas those living on estates are largely white working-class. Both, but particularly the former with their links with illegal drugs and so-called 'street crime', are designated by the police as 'high crime areas'. The police concentrate especially on the inner-cities, which through certain policing tactics and practices has given rise to much criticism of racial harassment, directed particularly against African-Caribbean youth. Keith (1993), for example, provides a provocative and detailed analysis of the relationship between the police and black communities in multi-racial inner-city areas in general and with reference to particular case studies.

What we shall illustrate with the following instances is a progression from inner-city to estate disorder. But alongside this there has been a definite development in the form of policing. Our focus will, again, be on certain watersheds and we also examine the more recent situation of estate disorders during 1991 and 1992. We start by briefly examining the riot in St Pauls, Bristol in 1980 and then move on to what is generally taken to be the first post-war inner-city riot, namely that in Brixton, south-east London, in 1981. We then look at Brixton 4 years later, in 1985, and in that year too, Tottenham in north London.

St. Pauls, Bristol - April 1980

The disorders in Brixton during 1981 are popularly taken to be the first outbreak of inner-city rioting of this type in post-war Britain. In fact, they were preceded by those

6. For clarification of the distinction between **inner-city**, **inner-city housing estate**, and **peripheral estate** to which we refer later in this section, see the glossary.

at St Pauls, Bristol, in 1980, which were taken at the time to be a 'one-off' case. It is worth mentioning St. Pauls briefly in order to put the later events in context.

On 2 April 1980 a number of police officers mounted a raid for drugs and the unlicenced sale of alcohol on the 'Black and White' cafe in Grosvenor Road on the 'front line' of St Pauls. It was realised at the time that such a move could potentially cause problems for public order, and there was an additional police presence with dogs. However, there had been no prior consultation between those planning the raid and the community liaison police officer who could be assumed to be more aware of community sensitivity, nor was the local Home Beat Officer present when the raid took place. In the event, the owner of the cafe was arrested, and a quantity of alcohol seized. A crowd gathered and at some stage stones were thrown at the police and a police car was set al.ight. Joshua and Wallace suggest that 'in scale and intensity, the violence (that followed) was then without precedent in post-war Britain' (1983:4). The police made a 'strategic withdrawal' for PSU reinforcements under the mutual aid scheme, resulting in St Pauls being an effective 'no-go' area for 4 hours during which time considerable damage was caused through burning and looting. A year later, in Brixton, much the same pattern was to recur.

Brixton, London - April 1981

Our main source for the events in Brixton, and for the subsequent recommendations for organisational and technological change in public order policing, is a report by Lord Scarman (1986) which was published in November 1981. The Scarman Inquiry was appointed by the Home Secretary with a specific brief to investigate the 'serious disorder' which occurred in Brixton (in south-east London) between 10 and 12 April 1981. In fact, the occurrences on the dates which Scarman referred to as 'scenes of violence and disorder ... the like of which had not previously been seen this century in Britain' (1986: sec.1 para. 2) can only be understood in the light of the events of a week earlier.

On 3 April, there was a police raid for drugs on a number of premises in the Railton Road 'front line' area of Brixton, resulting in 22 arrests. Three days later, on 6 April, the police mounted a concerted operation entitled 'SWAMP '81' in that area against 'street crime'. The internal police instructions stated that:

> The purpose of this Operation is to flood identified areas on 'L' District to detect and arrest burglars and robbers. The essence of the exercise is therefore to ensure that all officers remain on the streets and success will depend on a concentrated effort of 'stops', based on powers of surveillance and suspicion proceeded by persistent and astute questioning (Scarman, 1986:4.39).

This entailed the employment of 112 plain-clothed officers, and resulted in 943 people being stopped (over half of whom were black and more than two-thirds under 21 years of age) of whom 118 were arrested, but only one for the offence of 'robbery' (Scarman, 1986:4.40).

The 'SWAMP' operation utilised sec. 66 of the Metropolitan Police Act (1839), which permitted 'stop and search' on grounds of reasonable suspicion that the person so stopped was in possession of unlawfully obtained property. The resentment on the part of black youths, especially of the use of this offensive as discriminatory and harassing strategy, was as intense as it was, partly at least, because of the ill-feeling aroused by another widely discredited 'offensive strategy' adopted by the police, which by then had only recently been repealed, namely that known as 'SUS'. This entailed powers of arrest under sec.4 of the Vagrancy Act (1824) on grounds of 'SUS'picion concerning 'loitering with intent to commit an arrestable offence'.

The primary flashpoint for the initial disorder on 10 April, however, was a police attempt to assist a black youth who had been stabbed which after a police pursuit, was interpreted by 'the crowd' as hostile. This eventually resulted in confrontation between about 40 police officers, some of whom were in 'riot gear', and police dogs and about 100 youths. Eventually, the police gained control of the situation and instituted 'high-profile' saturation policing. Despite the general situation of extreme tension, a policy decision was taken to continue operation 'SWAMP'. This, according to Scarman, was a serious mistake:

> the police must carry some responsibility for the outbreak of disorder. First, they were partly to blame for the breakdown in community relations. Secondly, there were instances of harassment and racial prejudice among junior officers on the streets of Brixton which gave credibility and substance to the arguments of the police's critics. Thirdly, there was the failure to adjust policies and methods to meet the needs of policing a multi-racial society (Scarman, 1986:4.97).

Further, he argued that the police were operating in a situation of popular alienation:

> Tension between the police and black youths was, and remains, a fact of life in Brixton. Young black people, as well as many local people of all ages and colours, lacked confidence in the police. The worst construction was frequently put upon police action, even when it was lawful, appropriate and sensible (1986:3.23).

On 11 April two 'SWAMP' officers searched a 'mini-cab' (taxi) driver (outside the S and M Car Hire office) after seeing him push something into his socks which they suspected was drugs. In fact it was a number of bank notes but the police went on to search the car. A crowd of about 30 people had gathered, which soon increased to about 150. One person was arrested for 'destruction', and back-up was called for after an officer was erroneously thought to have been stabbed. The crowd then started a missile attack. Eventually a police van was set al.ight and there were a number of police charges. Looting took place and for the first time in mainland Britain petrol bombs were thrown at the police. In 4 days the toll of the rioting amounted to 145 buildings damaged; 207 vehicles damaged or destroyed; and 450 people reported injured. In addition 354 people were arrested and 7,300 police officers deployed (D. Waddington, 1992:82).

At one stage the police cordoned off the area for 3 hours until reinforcements arrived. Again, the police were severely criticised for allowing a 'no-go' area to develop, but Scarman felt their strategy 'first to contain the disorder (until a build-up of reinforcements) and then to quell it by an advance from both ends of Railton Road ... (leaving) means of escape open to the crowds ...' (1986:4.93) was reasonable in the circumstances. However, he made a number of points about the lack of protective clothing or helmets to cope with petrol bombs, problems affecting police radio communications on the ground and general co-ordination. More particularly, as regards the deployment of riot shields, he made a point which (as will be suggested later) was fairly clear even at the Lewisham disorder in 1977, namely that:

> the use of protective shields carried by officers can serve to attract missiles from a crowd and encourage officers to adopt a largely defensive posture, with the result that lines of police officers behind shields effectively become 'Aunt Sallies' for the crowd to aim at. There may be scope for the adoption by the police of a more positive, interventionist role in quelling disorder in order to speed dispersal and reduce casualties (1986:5.73).

As we suggest later, it was not really until after Tottenham 1985 that tactical shield units were put on a more mobile and offensive footing.

It is popularly thought that it was the Scarman Report that led to changes in the method of policing; indeed Scarman's recommendations are sometimes used to justify changes in police tactics in a generally more reactive direction. Such changes, it can however be argued, were against the spirit of Scarman, who concluded that:

> there should, I suggest, be no change in the basic approach of the British police to policing public disorder. It would be tragic if attempts, central to the thrust of my Report, to bring the police and the public closer together, were to be accompanied by changes in the manner of policing disorder which served only to distance the police further from the public (1986:5.74).

It will, in contrast however, be suggested later by one of those we interviewed that the changes in public order training and technology deployment that have been made post-Scarman have in fact reduced the scope the police have to de-escalate potentially riotous situations.

Scarman did make several recommendations for change, including that there should be:

> (i) means of ensuring that available police units are rapidly reinforced in the event of disorder by sufficient, properly trained and equipped officers. Effective reinforcement arrangements both within and between police forces are particularly important, because the traditional British approach to handling disorder requires, if it is to be effective, the presence of large numbers of officers;

(ii) increased training of officers, at both junior and command levels, in the handling of disorder. I have recommended earlier the adoption of common minimum standards and programmes for such training ...;

(iii) more effective protective equipment for officers - including better helmets, flame-proof clothing and, perhaps, lighter shields;

(iv) vehicles for transporting police officers which have some form of protection from missiles;

(v) improved arrangements for communication between officers involved in handling disorder and their operational commanders;

(vi) a review of police tactics for the handling of disorders (1986:5.72).

However, as Northam points out, much of this had already been decided by 'ACPO ... (who) held their crucial session on public order two months before the Scarman Report was published' (1988:51).

The following July similar riots took place 'in more than 30 towns and cities' (Bunyan, 1981:153) throughout Britain including Southall, west London, Brixton again, Mosside, Manchester and Toxteth, Liverpool. The latter is especially noteworthy in that it was the first time the Home Secretary authorised the use of CS gas on the British mainland; 59 rounds plus 15 gas grenades were fired into the crowd. One person was killed and another seriously injured by police 'vehicle tactics' (Vogler, 1991:133-135).

1985: 'The worst scenes of rioting yet'

In 1985 there was a resurgence of serious inner-city unrest in England and Wales in which four people died. This occurred in areas of London in the south-east of England, Bristol in the west, Birmingham in the Midlands and Liverpool in the north. It should not be thought, however, that the period between 1981 and 1985 was one of calm. Benyon and Solomos (1987a) refer to a number of 'mini-riots' which occurred in 1982, 1983 and 1984.

In the following section, we will concentrate on two specific instances of disorder during 1985, in Brixton, an inner-city area of south-east London as we have already mentioned, and in Tottenham in north London. Rioting here was confined to an inner-city apartment housing complex called the Broadwater Farm estate. Both incidents were preceded by police raids on private houses, the first resulting in the police shooting and paralysis of an innocent woman and the second in the fatal heart attack of another. Deaths also occurred in each of the two subsequent riots: one of a freelance photographer and the other of a police officer. The 1985 riots, and especially that at Tottenham, were particularly significant for bringing about changes in the police Command and Control structures and PSU tactics and the introduction of evidence-gathering techniques. Our main sources for the following are a report compiled by the Greater London Council Police Committee Support Unit on the events at Brixton and Tottenham (1986), the Metropolitan Police Commissioner's Public Order Review, also dealing with both incidents (Newman, 1986), and the report of the Gifford Inquiry into the disturbances at the Broadwater Farm estate (1986).

Brixton, London - September 1985

The riot in Brixton took place within three weeks of serious disorder in the Handsworth area of Birmingham in the Midlands. This lasted two days, during which two people were killed by fire. The 'flashpoint' for this disorder seems to have been the attempted arrest by the police of a black man for a suspected traffic offence, which took place against a background of allegations against the police of racial harassment and of a police 'clamp-down' on drugs in the area. The spark for the Brixton unrest in September was the shooting of an African-Caribbean woman, Mrs Cherry Groce, early on the morning of 28 September during an armed police raid on her house in search of her son who was wanted for a firearms offence. By afternoon a large number of people had gathered outside her house, and as Newman reports: 'Tension was gradually building up in the community' (1986:1.3.5). However, an attempt to defuse the situation was made at an early stage, the report continues, but 'for a variety of reasons, difficulty was experienced in contacting some community leaders' (1986:1.3.5). When the Chairperson of the Lambeth Community Police Consultative Group together with the Deputy Liaison Officer did arrive at the house, they were clearly unwelcome. At 2.45pm it was decided to open the operations room at New Scotland Yard to 'co-ordinate police deployment and reserves' (GLC, 1986:123). About two hours after this, a peaceful deputation of approximately 50 people set out for Brixton Police Station to seek information about the shooting and were joined by others on the way. A few people went inside the police station but came out dissatisfied, suggesting that they 'were treated with disrespect'. It was this specifically that the GLC Support Committee suggest was the probable cause of the escalation of the situation (GLC, 1986:123). Shortly thereafter some youths attempted to assault the police station from the rear, but were repelled by police officers. However, the former then started throwing bricks and bottles. A crowd of about 200 people had now gathered. At one stage a petrol bomb was thrown at two members of the Lambeth Community Police Consultative Group who were attempting to quieten things down. By about 6pm between 3,000 and 4,000 youths had gathered outside the police station and two more petrol bombs were thrown, and bricks were thrown through the police station windows.

The choice for the police at this stage, according to the Metropolitan Police Review, was to either:

> (i)'allow the crowd to riot in the vicinity of the police station, with the likelihood that the station would be invaded and the occupants injured'; or
>
> (ii)'attempt to disperse the crowd. The risk of this option was that the police were not immediately available in sufficient numbers to effect a controlled

dispersal of the crowd' (Newman, 1986:1.4.5).

In hindsight, however, we suggest that there was a third option, drawing on the later Tottenham Police Station experience, of mounting another attempt at de-escalation through negotiation, involving the Groce family and community leaders, accompanied by a declaration of apology. However, in the event the second option was chosen: 'the crowd was dispersed from the police station by officers in protective clothing, carrying shields' (Newman, 1986:1.4.6). The effect of this, as reported by the GLC Police Support Committee, 'was to split the crowd into 3 directions ... passing cars were stoned, stationary cars were set al.ight, shops were smashed and stripped'. The police then took action to 'seal off' the Brixton area by road, rail and underground.

The number of police deployed had increased throughout the day:
3.00pm: 27 District Support Units (DSUs) (324 officers)
4.30pm: 880 officers including 35 DSUs from outside the district
5.00pm: 1,212 officers, including dog handlers, mounted police and 72 outside DSUs
6.00pm: 1,556 officers
7.00pm: 1,793 officers, including 272 from the SPG (GLC, 1986:123-124).

By this time the police were engaged in running battles with approximately 1,500 youths, and the GLC report details a number of serious instances of excessive force being used by some officers (GLC, 1986:124-125). However, during the night, there was relative calm and the next day the police maintained a high profile, continuing the area 'containment' and undertaking a massive 'street clearing operation' (Newman, 1986:1.52). There were more riotous incidents that evening.

On the third day, the main policing strategy was that of 'intensive foot patrolling by neighbourhood officers' (Newman, 1986:1.5.6) together with CID teams mounting a post-riot investigation of offences including 'murder, rape, arson and robbery' (Newman, 1986:1.6.2).

Tottenham, London - October 1985

i. Background to the rioting

A week after the shooting of Mr Cherry Groce and the rioting in Brixton, another African-Caribbean woman, Mrs Cynthia Jarrett, collapsed and died in Tottenham (in north London) on 5 October during a police search of her home for stolen goods. Members of her family who were present at the time alleged that she fell after being pushed by a police officer. The police, however, suggested that she collapsed after one of her sons 'had arrived home and objected to the search' (GLC, 1986:126). The Coroners Court decided two months later that Mrs Jarrett was accidentally pushed by a police officer 'causing her to fall and contributing to her death through hypertensive heart disease' (Gifford, 1986:4.2).

The motive for the search operation appears to have been rather questionable. Another of Mrs Jarrett's sons, Floyd, who was well known in the community as a founding member of the Broadwater Farm Youth Association, was stopped earlier that day by police whilst driving a car and accused at first of a traffic offence, then of having stolen the car. Eventually he was charged with assaulting one of the police officers and taken to Tottenham Police Station.[7] Whilst he was there, four police officers took Floyd's keys to his mother's house (where he was not living) without booking them out with the station's custody officer, and used them to enter the house without knocking (Gifford, 1986:4.29 and 4.30). The Jarrett family insisted that no search warrant was produced, despite their request, whereas the police officers contested that one had been read out, although they admitted that no copy was left (Gifford, 1986:4.31). Floyd was held in custody for 51/2 hours and then released, only then being told on his way out of the police station that his mother 'had had a stroke'. He did not know at that point that her house had been searched (Gifford, 1986:4.55).

During the evening a number of community leaders visited the Jarrett family to express their condolences. The Chief Officer at Tottenham Police Station at that time, Chief Superintendent Stainsby, also visited the house, informing the family that 'an independent investigation would be carried out by an officer from another force' (Gifford, 1986:4.56). However, no answers were forthcoming about the way the officers had entered the house, about the search warrant, nor about whether or not they had been suspended from duty. The police issued no apology for what had happened, nor did they suspend the officers involved pending an inquiry; but the Metropolitan Police did issue an official statement that evening saying:

> Mrs. Jarrett was initially very co-operative. But towards the end of the search, another of her sons arrived home and began strongly objecting to the presence of the police. She collapsed and the officers were physically shoved out of the house (Gifford, 1986:4.57).

Later that evening, some members of the Jarrett family and community leaders, including Floyd who was now on bail, went to Tottenham Police Station to seek more information about the search and to see the warrant, but the original was not produced for them. Whilst they were there, a crowd of about 30 people assembled outside the police station and four of its windows were smashed. However, the crowd dispersed after appeals by the Jarretts and community leaders (Newman, 1986:2.3.6). One member of the West Indian Leadership Council stated, before leaving the police station, that 'This is no longer a family matter. It's gone beyond that. It has now become a community matter, and I think that it is important that somebody of some importance makes a statement' (Gifford, 1986:4.59).

In the early afternoon of the following day (6 October) a meeting arranged by the police took place at Tottenham Police Station. This meeting included members of the local

---

7. Floyd Jarrett was acquitted of this when his case came before the Court in December 1985 and was awarded costs against the police (Benyon and Solomos, 1987a:9).

Community Relations Council, the Police Liaison Committee, a community relations Officer and members of the Jarrett family. Demands were again made for further information regarding the search warrant and also for the suspension of the officers concerned, but to no avail beyond the assurance that matters would be investigated by the Police Complaints Authority under the Chief Constable of Essex. The Gifford Inquiry concluded that 'the failure to suspend was short sighted and insensitive' (Gifford, 1986:5.6). Certainly this did not help an increasingly threatening situation. During the meeting a crowd of over 100 people gathered outside the police station. However, no attempt was made to disperse them despite some stone throwing, and the police response was basically to let the demonstrators 'burn themselves out and vent their anger' (GLC, 1986:127), although there were DSUs deployed at the rear of the police station and in the neighbouring streets. The Gifford Inquiry praised the police response at this stage as 'restrained and sensible. They ... policed the demonstration with a thin line of officers in ordinary uniform, with special units well out of sight' (Gifford, 1986:5.7). Eventually, after about 11/2 hours, the demonstrators dispersed quietly.

ii. Sparks and tinder

Tension was meanwhile mounting on Broadwater Farm estate. Later that afternoon, bottles and missiles were thrown at two Broadwater Farm home beat officers when on the estate. One officer was severely injured. In the early evening, when the duty police inspector drove into the estate, he was injured by a beer bottle which smashed his car window. Shortly after this, police reserve units were released from duty by Area Control. The Metropolitan Police Review states: 'It appears that when the shifts were changing over, the progress of off-duty officers southwards through Tottenham *en route* to their home divisions was interpreted as a deployment to 'seal off' the estate' (Newman, 1986:2.3.24). Certainly, the GLC Report argues that 'at 6.30pm, groups of youths leaving the estate encountered police in riot gear. It is reported that they found police blocking the other exits from the estate (GLC, 1986:127). According to the Metropolitan Police Review, two DSUs went into the estate with 'protected transit vans' and one of these was attacked by a crowd of about 200 youths wielding 'bricks, petrol bombs and machetes' (Newman, 1986:2.4.2). In contrast, the Gifford Inquiry found that three police vans entered the estate with police dressed in riot gear, but the lead van was stopped by youths banging on its side with their hands. The officers then blocked one of the exits from the estate. It was only after this that youths overturned cars and set them alight as barricades (Gifford, 1986:5.33-5.34). In any event, police reserves were called and shield units were stationed at each of the four exits from the estate.

iii. The wider context

We should mention here some more general factors which account not only for the generally inflammatory atmosphere of the estate but also for the large police presence in the area. First, a feeling of tension and heightened expectation of trouble was to be expected given not only the recent violence in Brixton, but also the fact that during the week there had been rioting in Liverpool and also Peckham in south London. Secondly, there had been a recent police 'clamp down' on drugs on the estate, with a large-scale stop-and-search operation being mounted on 1 October. Further, it seems that this operation was mounted by the police without following the fairly well-established community consultative procedures (Gifford, 1986:3.63-3.65). Thirdly, there had been a build-up of police reserves in the area from 5 October, the day on which Mrs Jarrett died, and the District Control room was made operational as trouble had been forecast at two large shopping centres nearby (Newman, 1986:2.2.23; 2.3.1).

iv. Policing strategy and tactics

The contingency plan in existence at the time for policing a potentially riotous situation on the Broadwater Farm estate seems to have provided only one option, that of early resolution. The Metropolitan Police Review states this clearly:

> With the benefit of hindsight, police might have forestalled the disorder by the immediate deployment of uniformed officers on the estate. This would have been in accordance with the plans for such a contingency ... (which) was designed to provide for a containment of the rioters, the occupation of high vantage points such as walkways and footbridges, particularly there where police were vulnerable to missile attack, and the dispersal of rioters towards the west, away from the estate (Newman, 1986:2.3.21-2.3.22).

However, this option was not taken, and police shield units were deployed at each of the four exits to the estate:

> By about 8pm a pattern of behaviour by the rioters had been established. About 200-300 youths would emerge from under the tower blocks and attack one or more of the barricades locations with bricks and petrol bombs ... rioters (then) withdrew to re-arm, and would then attack again, either in the same location or at one of the other main locations (Newman, 1986:2.4.5).

The Gifford Report criticised the police tactic of 'For hour after hour ... (standing) in their lines across the road, behind their long perspex shields, passively fending off the missiles that came at them' (Gifford, 1986:5.52). The GLC Report suggested that it was questionable whether the police should have moved forward and attempted to breach the barricades, or withdrawn out of the range of the missiles. Whichever was chosen, however, they argued that: 'They should not have stood still' (GLC, 1986:128), which is what they did. There were some short shield 'snatch squad' sorties, but generally these withdrew rapidly without back-up or arrests (Gifford, 1986:5.52).

The Metropolitan Police Review pointed to 'a number of weaknesses in the police operation' (Newman, 1986:2.5.2) and the Gifford Inquiry even stated that 'we are in a state of some uncertainty as to who, if anyone, took overall charge of the night's operations' (Gifford, 1986:5.58). This echoed the GLC Report which suggested that:

> Police command was confused and at times broke down. There appears to have been two Commanders on the ground ... At times open disagreements

> between senior officers over deployment and tactics (whether or not to go onto the estate) were heard over the radio ... The men arriving from other parts of London did not know where they were or what they were meant to be doing (GLC, 1986:128).

v. The riot stage

During the riot, the police were attacked with missiles, including 'bricks, paving stones, bottles, tins, knives and petrol bombs' (GLC, 1986:128) and also, for the first time in an inner-city riot situation, guns. Two shotguns, a .22 and a .38, were fired in three incidents, injuring seven police officers. At some stage, and there was some debate as to whether this was before or after a shotgun incident, Sir Kenneth Newman, the Metropolitan Police Commissioner, authorised the deployment of baton rounds (plastic bullets) and CS gas. Members of D11, the firearms unit, were present (GLC, 1986:127) but none of these were used. Some time later, it was learned that a police officer, Keith Blakelock, had been murdered on the estate by people with a variety of weapons including machetes. Apparently he had entered the estate with a few other officers with the aim of providing assistance to the fire brigade (Newman, 1986:2.4.11). By about 4.30am, things had quietened down on the estate, and the police moved in force.

We feel that it is important to remember that not all people in the vicinity of a riot, or forming part of the crowd, such as that on the Broadwater Farm estate, are actively involved in the riot. As mentioned above, the nucleus, according to the Metropolitan Police Review, seemed to consist of 200-300 youths. The Gifford Inquiry report gives a description of the range of activity taking place on the estate during the course of the riot:

> During the evening of 6th October there were many people involved in many different kinds of action. Some were trying to use their position as community leaders to be intermediaries with the authorities, as far as they were able to be. Some were looking on, and we have heard many stories of residents coming out on to the balconies and chatting to the youths who were more directly involved. Some threw missiles of various descriptions, and set cars on fire. Some ... were prepared to commit arson which endangered life. Some were prepared to endanger life by shooting (Gifford, 1986:5.79).

vi. The post-riot situation

There were only a few arrests made during the riot period, but Broadwater Farm offers the first example of large-scale evidence gathering by the police in such situations As the Gifford Inquiry reported, the police had 'taken a huge number of photographs on the night', and these were used in a large-scale house-to-house CID (Criminal Investigation Department) investigation after the riot. Indeed, for three months after the riot, police were deployed on a massive scale. The post-riot 'aid requirements' to the Broadwater Farm estate and surrounding area for this period vary from:

7 October	1,409 officers
8 and 9 October	3,732 officers

to:	10-14 October (the period of the main arrests)	9,165 officers
	15-18 October	3,044 officers

declining to about 3,000 in the week of 12-18 November, about 1,700 per week between 19 November and 9 December, and 1,015 in the week of 10-16 December (Gifford, 1986:6.43). By the end of May 1986 the police had made 359 arrests, and 162 people had been charged for offences arising from the 6 October riot (Gifford, 1986:6.14).[8]

Peripheral estate disorder

So far, in respect of urban disorder, we have been primarily concerned with inner-city areas, a significant proportion of whose populations are from the ethnic minorities, although white youths were involved in the rioting. 'Peripheral estates' to which we now turn are predominantly white. Some show none of the poverty and general deprivation of inner-city areas, although others are perhaps even more marginalised, being located in the wasteland of cities without jobs or sense of community or recreational facilities. B. Campbell (1993) gives an interesting insight into these peripheral estate disturbances. She places the events in Tyneside, Oxford and Cardiff in the context of the residents' experience of economic recession which she sees as having led to a crisis of gender and generation in these areas, reflected in political tensions dramatically manifested in conflict between the police and local young men.

Within one week in August/September 1991 riots occurred in three cities. A *Guardian* newspaper editorial at the time, whilst recognising 'common themes' such as their all involving 'the poorer parts of relatively affluent cities, ... young people ... (and) peace-keeping police operations turning into violent confrontations' (*The Guardian*, 4 September 1991), stressed the need for an understanding of the specific features rather than sweeping generalisations:

> Each confrontation was in a different city. Each was triggered by different incidents: "hotting" (high speed stunts in stolen cars) in Oxford, black-out looting in Birmingham, and an ethnic dispute in Cardiff. Each police response was different: pre-emptive in Oxford, preventive in Cardiff, and proactive in Birmingham (*The Guardian*, 4 September 1991).

A fourth riot occurred a week after this on the Meadow Well estate in North Shields, north-east England. If Handsworth in Birmingham (referred to above as preceding the 1985 Brixton unrest) may be termed a classic 'inner-city' area, Meadow Well may be taken as a classic 'peripheral estate'. From 1991 on, the latter typifies the predominant form and location of urban riotous behaviour. It is to these four main incidents in 1991 that we now briefly turn, before examining the spread of peripheral estate unrest in

8. A number of people were also arrested and put on trial for the murder of PC Blakelock. Three of these were eventually found guilty and sentenced; however, disquiet remained about the nature of the CID enquiries following the event and the verdicts were subsequently quashed on appeal.

1992. Our sources for this section are mainly contemporary newspaper articles, D. Waddington (1992) and B. Campbell (1993), very little other work having been published on these developments to date.

i. Handsworth, Birmingham - 1991

It is questionable to what extent the disturbances in Handsworth on 2 September 1991 were planned, or as D. Waddington suggests, simply 'opportunistic' (1992:208). In any event, they occurred during a three-hour electricity power failure following a fire at a timber warehouse adjoining an electricity sub-station. The disorder mainly consisted of 'more than 100 youths rampaging through the streets' (Myers, 1991) and the looting of about 15 shops. The police were severely criticised by shopkeepers for the lengthy delay before they entered the area, which took place only after power had been reconnected. David Baker, Acting Assistant Chief Constable for the West Midlands Constabulary stated that he thought 'the police responded as quickly as possible, bearing in mind the area was totally blacked out' (Myers, 1991). D. Waddington, however, argues that it took about six hours from the start of the 'black-out' for the streets to be cleared (1992:208). According to Myers (1991) 'police admitted they took more than two hours to make a 'strategic decision' to wait until they had amassed detachments of 200 men in riot gear before moving in'.

That opportunistic looting should occur in such a deprived inner-city area, also given its previous history, and immediately following outbreaks of rioting elsewhere in England and Wales was hardly unpredictable. What was apparently surprising, however, was the geographical location of the rioting and the fact that the participants in the riots that occurred before and after Handsworth, came from peripheral estates.

ii. Ely, Cardiff - 1991

Four nights of rioting involving up to 500 youths throwing milk-bottles, stones and petrol bombs, with an air-rifle being fired (Katz, 1991), occurred on the Ely estate in south Wales on 30 August 1991. Ely is a relatively deprived area on the periphery of Cardiff, the capital, and consists largely of dilapidated social housing with a white working-class population. It is generally thought that the unrest was racially motivated, being initially directed against a 'Pakistani' grocer, and only later the police. The trigger seems to have been either a court injunction obtained by the grocer to prevent his white Welsh newsagent neighbour from selling groceries, or from the heavy-handed treatment he gave to some young shoplifters.

Events in Ely, and certainly the underlying tension, may be explained by the extent of marginalisation and unemployment and the sense of hopelessness of the residents as described by the local Member of Parliament: 'These people have simply lost faith that the future is going to be any better' (quoted in D. Waddington, 1992:211). However, the riots that followed on the Blackbird Leys estate in Oxford had a different background.

iii. Blackbird Leys, Oxford - 1991

The *Guardian* editorial of 4 September 1991 mentioned above referred to the incidents at Blackbird Leys, Handsworth and Ely as having a common factor in terms of their relative deprivation compared with neighbouring (or in the case of Handsworth, surrounding) cities. However, this needs some qualification in respect of society at large. The Blackbird Leys estate, for example, would certainly appear to be relatively deprived when compared with Oxford, but relatively prosperous when compared with peripheral estates such as Ely. Unlike Ely, for example, Blackbird Leys is more or less evenly divided between social and private housing (Carvel, 1991). Further, unemployment was not especially high in 1991, being only 9% (D. Waddington, 1992:208).

It would seem that the primary cause of the rioting was a police clamp-down on a form of 'vehicle crime' that they acknowledged had constituted a consistent pattern and which they had kept under 'constant surveillance' for the previous 18 months. This involved the theft of high-performance cars by youths from the estate which were then used in 'pre-arranged driving displays' on certain roads in the estate (Carvel, 1991). These so-called 'hotting' displays were attended by crowds of up to 250 onlookers. On 29 August the police made 20 arrests and beginning on 30 August there followed four nights of rioting, with bricks and petrol bombs being thrown at the police. This was, according to D. Waddington, in 'response to heavy handed policing of improvised youth activity' (1992:208). The criticisms of the police related, first, to the timing of the sudden police operation after such a long period of simply observing such events and, secondly, to the use of excessive force against 'innocent bystanders' (Bunting, 1991).

iv. The Meadow Well estate, North Shields - 1991

Among the three peripheral estates mentioned so far, Meadow Well (in north-east England) stands out as 'one of Britain's most deprived estates' (Wainwright, 1991a) with people existing in desperate poverty and in what a *Guardian* editorial described as a 'cesspit of unemployment' (11 September 1991). A Meadow Well Action Group survey conducted in 1991 found an unemployment rate of 87% (Wainwright, 1991b). Others have suggested figures ranging from over 40% for the estate as a whole (D. Waddington, 1992:210) to 85% for 'young people' on the estate (*The Guardian* editorial, 12 September 1991).

The flashpoint, within the context of a 'tough approach of Northumbria police to car thefts and ram raiding' (Wainwright, 1991a), was the death of two youths from Meadow Well in a stolen car as a result of a police chase. This led to Meadow Well undergoing a '5-hour rampage of looting and arson' (Wainwright, 1991a) by up to 400 youths on 10 September, during which:

> A school, community centre and health clinic were all firebombed. A long row of shops, several owned by (south-east) Asians, were looted of goods

> worth thousands of pounds, and then set al.ight. A firebomb on an electricity sub-station blacked out part of the estate. Trees were uprooted and strewn across roads (as barricades) (*The Guardian* editorial, 11 September 1991).

Sir Stanley Bailey, the Chief Constable for Northumbria, 'admitted that control had been lost at Meadow Well. He defended as perfectly proper in the circumstances the contain, not confront operation' (Wainwright, 1991a). The unrest in Meadow Well spread to a number of other peripheral estate and inner-city areas of Tyneside, resulting in two weeks of rioting in all.

1992: The problem continues

Within a period of just over two weeks during 1992 there were serious outbreaks of disorder on at least eight peripheral estates located in different parts of England, namely:

May 12-13:	Wood End estate, Coventry, Midlands
June 15-16:	Ragworth estate, Stockton-on-Tees, NE
July 1-6:	Ordsall estate, Salford, NW
July 6-9:	Marsh Farm estate, Luton, SE
July 16-18:	Hartcliffe estate, Bristol, SW
July 19-24:	Stoops estate, Burnley, NW
July 22-23:	Brackenhall estate, Huddersfield, NW
July 22-24:	Whalley Range estate, Blackburn, NW.

As in the previous year, the unrest involved mainly white working-class youth and occurred almost always as an immediate response to police curtailment of illegal 'leisure activities'. We will briefly mention some of these instances in more detail, before considering the implications for policing practice.

i. Wood End estate, Coventry - 1992

The immediate cause of two nights of disorder involving about 200 youths in battle with police, looting shops and firebombing a school on what has been described as this 'drab fifties estate' (Katz, 1992) on the outskirts of Coventry in the English Midlands, would seem to have been police action against youths riding 'off road' scrambler motorbikes in a dangerous manner all over the estate (Johnston, 1992a). Like the 'hotting' in Blackbird Leys, this had been occurring for some months previously. The trouble erupted shortly after a 'police task force' arrested three riders and confiscated their bikes, although there were also suggestions that the general tension on the estate was linked to 'a shortage of marijuana following a number of recent police raids' (Katz, 1992).

ii. Ordsall estate, Salford - 1992

The Ordsall estate, located in Salford in north-west England and consisting of 2,000 houses and apartments built between the 1960s and the 1970s (Foster, 1992), had also been the object of a prolonged police operation against young criminals. This was similar to that at Meadow Well, namely 'preventing the theft of vehicles and their subsequent use ... for ram raiding' (Bunyan, 1992). To this end, police had, for some time, 'been deployed in large numbers around the estate' at night (Bunyan, 1992). This would seem to have led to six nights of unrest involving a number of firebomb attacks on neighbourhood council, housing and unemployment offices and a restaurant and also police and firefighters being shot at with a handgun in three separate incidents, eventually leading to the deployment of armoured police vehicles.

iii. Hartcliffe estate, Bristol - 1992

Again as in Meadow Well the previous year, unrest occurred on the Hartcliffe estate in Bristol (south-west England) after the death of two 'joyriders'. In this instance, they were riding a stolen police motorcycle and died after being in collision with an unmarked police car containing two officers from the regional crime squad (who were then suspended pending an investigation by the Police Complaints Authority). For three nights there followed rioting, looting and firebombing by over 100 youths (Johnston, 1992b).

iv. Whalley Range estate, Blackburn - 1992

Unlike the above cases, the disorders on the Whalley Range estate involved clashes among about 800 Indian and Pakistani youths arising out of a fight between two youths over a girl. A cafe was stoned and set on fire, and on their intervening, the police became the target for violence (Oldfield, 1992).

The policing dilemma: strategy and tactics

What we have described in the above instances of peripheral estate disorders in 1991 and 1992 is not so much the problem for the police of dealing with a riotous situation when it occurs, or even with a post-riot situation - fairly standard procedures seem to have been developed for dealing with such eventualities - as that of controlling criminality without causing escalation in a potentially riotous situation. In a number of the examples we have outlined, there has been criminal activity which police have not only been aware of but have been monitoring and gathering evidence on for long periods of time, before mounting control operations. The police have been criticised on a number of these occasions, in retrospect, for waiting before taking action on the grounds that this may strengthen the resolve of those engaging in criminal 'leisure' activities. However, the dilemma posed by car theft and 'joyriding'- and it would seem that this has been behind a number of recent outbreaks of disorder on peripheral estates - is a very real one. This dilemma has been expressed by D. Waddington in reference to Blackbird Leys: 'Unable to effectively give chase to the hotters for fear of endangering the lives of pedestrians and motorists, the police had been forced to play a waiting game based on covert surveillance' (D. Waddington, 1992:209). A 19-year old male, speaking of the immediate cause rather than the underlying problems of marginalisation, unemployment, deprivation and hopelessness generally existing on these estates, described the battle of Wood End estate in Coventry as having occurred simply because: 'They (the police) came round in cars with two with a camera in the back and two in the front giving us the finger. We gave it back to them' (Katz, 1992).

A number of other urban riots have occurred since those we have detailed here. Indeed, rioting has just taken place in Manningham, an 'inner-city' area of Bradford in the north of England, at the time of writing (9-11 June 1995), by 'hundreds of youths' primarily of South Asian origin, involving the throwing of bricks and firebombs at police and reportedly causing more than £1 million of damage to property. The immediate cause, it would seem, was 'heavy-handed policing' of a minor disturbance; but there were deeper-seated contributing factors, ranging from an ongoing local conflict concerning prostitution in the area to an increasingly alienated and marginalised second-generation Pakistani youth (Wainwright, 1995).[9] Following to some extent what B. Campbell has described as a 'routine' (1993: xi), most disturbances in the more contemporary period have, however, not been on the same scale as those in 1991/2. It has, as one of our interviewees suggests (in the next chapter), been a continuing problem of policing minor skirmishes and of simply 'keeping the lid on'.

9. According to the 1991 census, 9.9% of Bradford's population are of Pakistani origin, and Bradford's South Asian communities number approximately 70,000 (Donegan, 1995; Wainwright, 1995).

Chapter III

Lessons Learned and the Rationale for Change: public order policing of the past, present and future

Introduction

For this part of our research we conducted a series of in-depth and wide-ranging interviews with senior public order police officers from 3 forces, the British Transport Police (BTP), Kent Constabulary and the Metropolitan Police Service. The views of our respondents are their own, and do not necessarily reflect those of their respective forces. They therefore remain anonymous, and we refer to them simply by their ranks. Indeed, as we shall see, some difference of opinion as to the best strategy and tactics is expressed occasionally by respondents within the same force. We do not suggest that the opinions expressed here are in any way representative of a body of 'police thought'; our respondents were not chosen to form a representative sample. What these views do represent, however, are various understandings of the way in which policing strategy and tactics are (and have been) developing, from a number of practitioners with varied professional experiences. Their responses reflect the debate which is going on within the police community as a whole.

The three forces themselves are involved in policing different situations with a potential for disorder, and to some extent the differences in the experience of our practitioners and the problems they face determine their approach to the issue of public order. For example, one of the main concerns of the BTP is to keep the 'crowd', be they football supporters, carnival-goers or demonstrators, mobile and unhindered. The Metropolitan Police on the other hand tend to be concerned with large-scale static crowds and policing the 'front line' of an inner-city area, while Kent Constabulary often faces smaller-scale 'skirmishes' such as drunken brawls and 'rave' parties (without being free of the threat of 'estate' or area disorder). Also, at the time of writing, Kent's docks area is the scene of animal rights protests.

Our focus in this chapter will be threefold. Initially we will be exploring the relationship between policing developments and specific 'watersheds' - events which triggered a reassessment of the practices then in use. We then consider such developments in terms of tactics (including policing methods, technology and training) and strategy (especially in relation to sensitivity to escalation/de-escalation). In this first strand, we will draw out references in the interviews to the main instances discussed in Chapter II above. In this and the following strands we will largely let the respondents speak for themselves. The second strand will concentrate on current 'state of the art' public order policing practices. Special mention will be made of the 'multi-agency' approach incorporating advance negotiation, contingency planning, evidence gathering before and during incidents, and pre-emption through tension indicators. We also focus on current technologies, training and command and control (**'Gold, Silver and Bronze'**) programmes. Finally, we indicate trends for the future.

1. The past

i) Grosvenor Square and after

Our respondents, almost without exception, saw the late 1960s Vietnam demonstrations as the first major watershed in post-war public order policing. It was suggested by a Superintendent in the Metropolitan Police that Grosvenor Square in 1968 introduced a whole new scenario: mass demonstrations with many people arriving from many directions and with a variety of perspectives on the issue about which they were protesting. Public order training had before this been reactive to problems. The respondent stressed that the police now try and anticipate problems so as to plan appropriate responses and reactions, mainly with an eye to historical precedent. Rather than suggesting that any one strategy or tactic had come to supplant others, the officer claimed that 'push and shove' tactics, cordons, the deployment of shields and protective clothing and evidence-gathering using photography and video-taping might be employed at different times and in different circumstances - the public order policing repertoire has merely been enhanced and diversified.

A Chief Superintendent in the Metropolitan Police agreed that Grosvenor Square was the first large-scale demonstration that had the potential for serious disorder. He added that the number of injuries suffered by the police had certainly caused concern and that, despite the somewhat light-hearted way in which the event had been presented, 'having pennies thrown at you was no joke!'. This respondent identified Red Lion Square (1974) as the next main watershed. Here the problem was that people could not leave the square once violence had erupted and a death resulted.[10] This brought a heightened awareness of the need for forward planning and better organisation of police resources.

A BTP Chief Inspector added that, in his view, between the 1960s and 1976 the style of policing has been individually physical, with much face-to-face contact, involving very little strategy and very little thought. Until the March 1968 Grosvenor Square disorder there was very little thought given to public order tactics. After this event a range of options (such as linked arms, open cordons and filter cordons) were discussed and planned, although training was at best rudimentary and often non-existent. He noted that the pervasive image of the police officer in England and Wales as very much an 'individual operator', led to such 'military-style' organisation being seen as a break with tradition and as somewhat aggressive when first deployed on the streets and witnessed on television.

Some respondents qualified this view of history. A Metropolitan Police Chief Superintendent saw the major 'sit down' protests of the Campaign for Nuclear Disarmament (CND) in the early 1960s as the first post-war public order problem. A Metropolitan Police Inspector specialising in public order policing agreed that the CND protests had been problematic for the police owing to the huge numbers of generally peaceful participants whose 'sit down' protests made the use of mass arrests on minor obstruction charges the focus of police activity. It was suggested that it was probably the age and experience of current police commanders which led to Grosvenor Square being seen as such a major watershed. There are few, if any, officers still in service who had first-hand experience of the police operations that were undertaken to deal with the outbreaks of racially-oriented street gang disorder in Notting Hill, London and in the city of Nottingham in 1958.

ii) 'Push and shove' and tactical developments

Most of the comments concerning the traditional method of public order policing by 'push and shove' were critical of the notion that it was some kind of 'gentlemanly sport' (see Chapter II), but also highlighted the resource implications of policing public order by sheer weight of numbers. The costs of this could, they suggested, not be met by current police budgets. It was noted by a Metropolitan Police Inspector that 'push and shove' tactics were something of a ritual in the industrial relations field but they should not be romanticised. Such tactics could, and did, result in serious injury. In this officer's opinion the relatively favourable image of 'push and shove' was merely nostalgia born of the violence of subsequent events. An Inspector in Kent concerned with public order training referred to his earliest experience during the 1972 miners' strike when it became clear that there was a great deal of danger involved in linking arms and pushing and shoving in a

10. Scarman, already referred to as leading inquiries into Brixton 1981 and Grunwick 1977, also conducted an inquiry into public order policing and the policing of this incident in Red Lion Square in which Kevin Gately died. As in Southall in 1979, there were suggestions (although unproven) that Gately's death resulted from a baton charge by the **SPG** at anti-fascist protesters (Rollo, 1980:179-181) (see **Police Support Units** in Glossary).

picket line situation involving the movement of heavy vehicles. Although it was clear from this period that more flexible and improved tactics had to be planned, little was done for some years. The officer went on to describe the variations on 'push and shove' that were developed and illustrated their potential value in certain situations today:

> There is still some use for the tactics that were developed then, for example the 'snooker ball wedge', which is a modified basic 'trudge and wedge' manoeuvre, appropriate for circumstances such as effecting VIP entry to meetings etc. Officers are still training for different types of cordon: the 'loose cordon' is basically a line of officers with spaces in between and, predictably a 'tight cordon' is achieved when those officers close up; the tight cordon can be developed into a 'linked cordon' where officers hold the belt of the officer in front and keep one arm free.

A Metropolitan Police Superintendent added that in the 1970s it was quite usual to 'throw' officers at crowds, in order to contain them by weight of numbers and impromptu organisation. By way of illustration a Metropolitan Chief Superintendent claimed that the tactic of 'push and shove' probably reached its peak at the Saltley Coke Depot picket (1972) and finally become discredited as a large-scale tactic at Orgreave (1984). Another respondent, an Inspector from the Metropolitan Police, noted that prior to and during this period the issue of cost was not much considered. The fact that on some days up to a quarter of the Metropolitan Police strength was deployed at the Grunwick picket inevitably cost a great deal of money and reduced the ability of the force to fulfil other service and crime-related roles.

iii) Notting Hill 1976 and Lewisham 1977

The next watershed, again identified by all our respondents, was the 1976 Notting Hill Carnival. It was suggested by a British Transport Police Inspector with special responsibility for public order that after the 1976 Notting Hill Carnival there was considerable pressure from the Police Federation and from within the Metropolitan Police generally for the introduction of protective gear. First, the reinforced helmet with a chin-strap was introduced, while still looking like the traditional helmet, and to some extent this remains a guiding principle. Later, a visor was also provided. There were local and individual innovations, with some equipment being quite literally bought from sports stockists, such as football shin pads and groin protectors. More recently, the Police Scientific Service has taken a lead in developing and evaluating such special equipment. Long shields also came in after Notting Hill in 1976, and were first deployed at the National Front march in Lewisham in August 1977 and again later that same month at the 1977 Carnival. At Lewisham, despite reassurances by the Home Secretary that the shields would only be used as a last resort, they were on hand and ready to be used. A Kent Inspector noted that the issue of supplementary protection, such as shin, elbow and knee guards and shoulder and groin protection was revived when many forces moved from the use of long to shorter shields - a tactical shift developed in response to strategic development to which we shall return later.

A Metropolitan Police Superintendent told us that 'the escalating ferocity of crowd behaviour, notably after Notting Hill 1976 with the dustbin lid versus milk bottle scenario was untenable', whilst a Metropolitan Police Inspector reinforced this with his personal view of developments:

> Notting Hill 1976 was typified by the 'dustbin lid' response. I was there; it came as a total surprise. Shields were developed after this event and the Special Patrol Group then embarked on a training programme, followed by divisional officers.

Outside the Metropolitan force, training with long shields started on a national basis after the 1976 Notting Hill Carnival. A Kent Inspector recalled that in 1976 training in long-shield tactics was started by the Metropolitan Police for the West Midlands Constabulary SPG, and at about the same time Kent Constabulary also started shield training.

iv) Brixton 1981 and Scarman

The Scarman Report into the Brixton disturbances is commonly taken to be the most significant watershed in the policing of public order, but our respondents suggested otherwise. They pointed to the experiences at Notting Hill and Lewisham as the main watershed, with Brixton 1981 bringing out the stark 'reality' of the implications of the changing strategy and tactics and the Scarman Report as the legitimation for what followed. According to the recollection of one Metropolitan Superintendent:

> Post-Lewisham 1977 had seen the new, reinforced helmet and 'cricket box' introduced and this little change generated something of a feeling of invincibility that was swiftly challenged as injuries mounted. Enhanced protection became a part of training after 1981. Shield training was, in this period, to give officers *protection*, not to equip them to *achieve* anything. I think that training started after the 1976 Notting Hill Carnival. I was part of the first Police Support Unit trained for shields and we were trained by the SPG. At Brixton in 1981 the limitations of this 'protection, then what?' approach were fully illuminated. Snatch squads were not fully tried and a lack of fire protection also rose on the agenda.

In the view of a public order Inspector from the Metropolitan Police, the Brixton disturbances had a major political impact on policing policy in general, and public order policing in particular, in deprived and/or multi-racial urban areas:

> At Notting Hill in 1976, the frustration of black youth was barely recognised. It was with Brixton and the national disturbances in 1981 when the hierarchy recognised previous failings. Although the police had started to address some

> of the issues before the troubles and subsequent publication of the Scarman Report, it was the Brixton Riot that acted as the catalyst and the Report that provided the political motivation.
>
> It would appear that there had been a warning for Brixton in terms of St Pauls, Bristol in 1980. There was however a brief spell of unfounded complacency, especially in the Metropolitan Police, and the lessons were not learned because it was felt that Bristol had too few officers available to deal with the scale of the disorder, had not made any plans for how to deal with it, and had been caught on the hop by an unfortunate and unpredicted set of circumstances. It was only after Brixton that the true problems of police ethnic relations became clear and questions were asked which demanded change.

A British Transport Police Inspector noted that, from his experience, police services in England and Wales were already generating and developing ideas for change prior to the presentation of the Scarman Report into the Brixton disorders. These preparations were somewhat at odds with the spirit of the report which, nonetheless, served to legitimate the changes in policing policy, strategy and tactics. Indeed, in the view of this respondent, the police inability to deal tactically with the disorder served as a determinant for changes in strategy and policy.

v) The miners' strike 1984/1985 and Tottenham 1985

Rather than constituting a watershed in respect of instituting change, the miners' strike of 1984/5 was generally considered to have occasioned the application of strategies and tactics that had already been developed. Tottenham in 1985, however, would seem to have led to a review of command and control, Police Support Units and shield formation and training, and to have highlighted the role of 'evidence gathering' as a means of bringing offenders to book while, ideally, reducing the need for direct and potentially escalatory physical confrontation.[11]

It was suggested, however, by a Kent Inspector, that the miners' strike of the mid-1980s constituted a major organisational watershed with the implementation of what had formerly been, essentially, an emergency civil defence plan to co-ordinate local police services in England and Wales into a more centrally directed resource: 'During the miners' strike Mutual Aid was a major organisational watershed innovation. On the tactical front, on the other hand, Broadwater Farm was the 'nail in the coffin' for the slow long shield, five man team'.

---

11. Any subsequent criminal investigation process, like that which followed the Broadwater Farm disorders for example, would of course have a significant negative impact on the possibility of a community returning to a 'state of normality' as suggested by the 'riot curves' discussed in Chapter I.

A Metropolitan Police Superintendent outlined the strategic and tactical implications of the events at Broadwater Farm:

> The death of a police officer at Tottenham raised the issue of how the senior officer on the ground was to control his teams. Given the nature of the environment, there had really been no equivalent experience to build on. The event generated a reconsideration of how to deal with disorder in highly complex space such as a 1960s or 1970s housing estate. Senior officers had not been exposed to this scenario before, and a judgement was made early on that it was anticipated that the disorder would focus on efforts to target property and looting in Wood Green (a relatively affluent shopping centre nearby). This did not turn out to be the case, the purpose was to resist 'invasion' of the estate.
>
> This experience led to the watershed in terms of tactics, namely the enhancement of training for senior officers, and also led to some incentive, as public order awareness and training offered some grounds for promotion as a unique demonstration of command skills.

It was observed by an experienced Metropolitan Police Inspector who had been involved in public order training that:

> The result of a policeman being killed at Tottenham, a tragic and almost unprecedented event, was that more officers were trained in the use of baton rounds and mobilisation policies were reviewed. Training was stepped up and training and equipment reviewed. The Hounslow Public Order Training Unit was further enhanced. The issue of possibly deploying motorised and protected water cannons was again raised (as had been following Brixton 1981) and shelved due to an appreciation of the limited mobility, water capacity and vulnerability of the machines in complex urban space. An additional concern was the 'political' sensitivity concerning the radical departure such deployment would mark from the traditional 'civilian' image of British policing which would threaten to compromise the potential to exercise a minimum force style of policing in the 'post riot' stage.

An Inspector in the Kent Police with responsibility for public order training drew attention to the changes in strategy and tactics that the experience of the **Broadwater Farm** disorder generated even outside London:

> One of the main newer pieces of public order equipment is the 'Kent shield', or Armadillo, introduced in 1988. This was developed post-1985 Broadwater Farm when five-man units carrying long shields to keep ground against a hostile crowd were deployed, simply standing and taking a pounding. This led to a decision that such a situation could not occur again and that 'offensive' action was required.

A British Transport Police Inspector stressed that Broadwater Farm brought out the potential for intelligence gathering in support of post-event criminal investigation, both to avoid direct engagement which might escalate the situation and jeopardise the short-term defusion of conflict and to continue to preserve the possibility of the law being enforced in circumstances which had made it impossible for officers to operate 'on the ground':

> Despite the increase in 'firepower', officers could still not get onto the (Broadwater Farm) estate due to the ferocity of the opposition and the confrontational methods deployed (remember that at one end of the estate a Chief Superintendent with little by way of resources and 'firepower' had ensured a relatively peaceable environment). The general experience, however, stressed the importance of evidence gathering where containment and early resolution had not been achieved. Such tactics involved the use of a camera with high light intensity and a tape recorder to record what was going on, by way of a commentary to link up with the photographs, in order to identify offenders and trace them later. This was first used at Broadwater Farm, but the ground had been prepared after Brixton 1981 when a senior Photographics Officer (civilian) had suggested the potential.

The evidence-gathering techniques explored at Broadwater Farm, however, had little or no impact on the eventual prosecutions, convictions, acquittals and successful appeals of those charged with the murder of PC Blakelock. This technology is now routinely employed at large-scale demonstrations, in the monitoring of sports ground disorders and the surveillance of targeted 'football hooligans', and has supported some successful prosecutions.

vi) Contemporary urban disorder

There was little specific comment on the form or the policing of major contemporary urban disorders, although references were made to the necessity for mobility, speed and firmness of reaction. This, we feel, is somewhat reminiscent of Smelser's demands for 'firm but fair action' criticised in Chapter I. It was suggested that instances of spontaneous disorder were fairly commonplace, if not on the increase owing to the marginalisation of certain populations, in most urban environments. With 'rapid response' the tendency was towards early resolution.

A Metropolitan Police Inspector gave his view of the change in public disorder over recent years, from primarily inner-city to marginalised estate, while suggesting that 'local' disturbances were frequent occurrences and necessitated the tactical deployment of officers to head off potential escalation:

> We have had a qualitative change in disorder, reflecting social change ... In London such localised disturbances are dealt with quickly by Territorial Support Groups who 'keep the lid on' by deploying suitably acclimatised

and equipped groups of officers to secure early resolution of events which may provide the focus for further disorder.

vii) The problem of escalation

Our respondents showed a unanimous awareness of and sensitivity towards the problem of mutual escalation and the precipitation of conflict by police action, even if differing in their individual recipes. They voiced a number of criticisms of the strategic or tactical policing of specific past events, although of course it must be realised that such statements reflect both hindsight and subsequent changes of practice. On this basis, a British Transport Police Inspector was forthright in his criticism of the neglect of the interests of legitimate demonstrators in favour of a concentration on controlling the 'active core':

> There did seem to be something of a lack of common sense deployed at **Lewisham**, as after the National Front contingent had left the police attempted to disperse the protesters aggressively, using the new style helmets and shields, despite the fact that they were largely untrained in their use. It was an ideal opportunity to disperse the crowd using persuasion by providing information that the target/trigger had departed, but this information was not given. Lewisham therefore was something of a turning point in terms of the police being equipped to mount a punitive response.

This officer went on to suggest that there remains a focus on the maintenance of order, at the possible expense of a better relationship between police and public, an attitude in part perhaps due to an awareness among some in the police of the increasingly complex and, for policing, problematic nature of contemporary society:[12]

> The police would generally like public order strategy, even post-1976, to be thought of as accommodation: 'we are only here to help you', but deep down the attitude has changed to: 'we will negotiate so far, but after that, we are the boss and we will tell you what is going to happen', asserting a new boundary into the ideal policy of assistance and accommodation. This is very much an 'iron fist in a velvet glove' approach. Once that boundary is crossed there is a danger of moving from a jaunt to a riot very rapidly due to the aggressive nature of the likely response. The image is now of a more caring cop on the street in view of the public, with reserves all tooled up and ready to go in the backstreets. Realistically one can not expect officers to wait for disorder to occur before donning equipment as this usually takes about ten minutes, hence the necessity of keeping the heavy material on standby so that senior officers know that the minute that they make a decision they are ready to implement it.

12. This is an issue that has also come to the fore in the recent differences in the policing of Animal Rights protests.

> The police see the public as getting badder and badder and generally more violent and less respectful of the law. They do not seem to realise that the way that the police deal with situations (such as the deployment of support units) leads to escalation. At least this appears to be the police subcultural view ... The police have to realise that they are there to facilitate the activities of the public in daily life. People will expect to be able to protest and demonstrate in daily life and demonstrations are a legitimate form of activity that should be accommodatable. At the **Poll Tax demonstration** the police successfully negotiated a minefield of potential flashpoints, but when the boundary set through negotiation was crossed, they reacted in a manner that rapidly escalated the situation. Communication by meeting leaders must be enhanced and ground rules set, yet when these are breached a sensitive and subtle response is required, not an immediate, punitive escalation, apparently designed to punish the public for not playing to the rules.

This respondent also drew attention to the possible drawbacks of adopting a strategic policy dependent on the tactical deployment of reasonably well equipped and trained officers in rapid response support units to 'crush problems in the bud':

> This standardisation (of post-Scarman 'common minimum standards' for public order training) came about as a product of inner city problems and many senior officers are arguing about whether it is the most effective and appropriate way of dealing with such problems. The Thames Valley for example, are worried that deployment of level 3 trained officers, their equipment and tactics, may not have been the most appropriate way of dealing with the **Blackbird Leys** disturbances. A senior officer argued that deployment of such standardised equipment and tactics may well have contributed to the escalation of disorder on the estate ... It can be argued that such standardisation is vital in mutual aid cases, so that officers understand what is going on wherever they are deployed, but it has a negative impact on unnecessary escalation in many actual and potential circumstances.

Drawing on his personal experience, this BTP officer illustrated the potential for resolving conflict in a manner which is different from the prevailing ethic in public order policing and, as we shall see shortly, is one that his police force has found to be essential in the face of geographical and resourcing realities:

> I can remember an incident during football traffic on the Underground system when about 600 boisterous supporters had to change trains in order to reach their destination. They were held up for a short time underground waiting for their train and became quite disorderly. Members of the public were caught up in the crowd and the situation became quite frightening. Territorial Support Units were sent to assist officers in restoring order. On the return journey, after the match, in order to prevent a repetition of the earlier events, police slowed down the flow of supporters and any resulting delay and frustration by easing

> the trains into the platform one carriage at a time and emptying it. The supporters had previously been advised by the train announcing system what was to happen. The result was that there was no disorder at the interchange and extra reserves of police and dogs were not required.

Whilst a Chief Superintendent in the Metropolitan Police noted that less serious public disorders might be dealt with by the deployment of resources aiming to achieve 'early resolution' through firm and confident action, he stressed that there was a tendency now to exercise greater initial restraint at larger-scale events:

> Shields are rarely used nowadays on public order events. Putting uniformed officers in the right place at the right time is much more effective. Intelligence is vital here in order to allow pre-planning and to ensure officers are on the ground targeting problem groups before disorder escalates.

The difference between this and the earlier approach may to some extent be explained by the differences between the policing tasks of the BTP and the more geographically focused concerns that typify the police forces in England and Wales. This different perspective on public order policing was highlighted by an Assistant Chief Constable in the BTP:

> The BTP has little use for, or officers to serve in, Police Support Units. The Force does have MSUs (Mobile Support Units), but these are smaller units which, while quite heavily equipped, at the moment (under review) do not have a major public order function, they are mainly to provide back-up to individual officers on duty should an individual or small group present a particular problem ... The force has made a conscious strategy not to obtain baton rounds or gas for example, because they do not fit with the force remit and they do not have the personnel reserves to accommodate the training necessary to maintain the required levels of proficiency in this field ... Should a 'wheel come off', or seem likely to do so, support from this style of policing would have to be sought through the **ACPO** Mutual Aid arrangements with home forces.

The philosophy of public order policing adopted by the BTP would seem to be a reflection of the fact that its officers will frequently be completely unsupported at first and heavily outnumbered by potential opponents in the event of conflict, for example on a train carrying football supporters. The Assistant Chief Constable believed that this necessarily led to the preservation of a policing tradition based less on confrontation than on accommodation and subtle modes of control:

> The force policy is to build up individual rapport with the public using small numbers of officers. For example, I sometimes go out with officers on football trains or somesuch and demonstrate that an officer can develop a rapport with a group of rowdies and maintain order with humour and banter. An individual officer can maintain order in most circumstances in that way by winning the support of the bulk of a crowd and shoring up informal social control.

This respondent gave an example of how such a policing style could be maintained even in the face of relatively large-scale disorder:

> In August 1984, during the **miners' strike** 102 pickets took part in the Port Talbot crane occupation. Building relationships in such a situation was vital. I made sure that we did not have the sort of officers who waved fivers (five pound notes) at the miners or the 'ASPMM'[13] attitude. All the protesters were negotiated down from the crane, and arrested and charged for this offence, all taken to police stations by lodges (units of the mineworkers union) and dealt with sensitively. I made it clear that officers who taunted miners would be formally disciplined, not merely rebuked. You have to leave room for those policed to retire with dignity so that wounds can heal and decent working relationships be restored in the aftermath.

Having provided this example, the Assistant Chief Constable was at pains to point out that the types of event dealt with by other police forces might be of a more spontaneous character. However, the issue of careful forward planning for events was also stressed:

> Strategic Planning, Preparation and Co-operation can reduce public disorder. ... This is, however, imperfect in some circumstances. In the **Poll Tax riot** police communication broke down. The left hand did not know what the right hand was doing. There were radio failures; failures of strategy; contingency planning and in terms of briefing and debriefing.

The implications of attempting to maintain a 'minimal force' public order policing tradition was also stressed by a Superintendent in the Metropolitan Police:

> One of my main concerns is that in the UK, tradition, culture and legislation can lead to the police being reactive with some events that we could have been more adequately proactive to. We have the potential to deal with events of public disorder very effectively (such as by using baton rounds) but in ways that are not acceptable, or desirable in British society. Order can be rigidly maintained, but at what cost?
>
> An example of 'sensitive' or discretionary policing would be where an officer at the Notting Hill carnival sees an individual smoking a joint (cannabis). The officer has a legal duty to do something (they have to do something to enforce the law), yet we know that attempting an arrest in those circumstances may well precipitate something worse, so no action is taken: there is always the potential to follow up afterwards. If an individual officer chooses this course of action, they cannot be criticised in law.

This respondent went on to point out what he saw as the problem which resulted from having gradually adopted equipment that was, on the face of it, purely defensive rather than proactive in potential:

---

13. 'Arthur Scargill Pays My Mortgage' - a taunt allegedly directed at increasingly impoverished miners by police officers benefiting from overtime payments while attending the 1984/5 coal dispute.

> I do not know the precise origins of shield tactics, but even the Roman Army knew that such a slow moving wall of shields drew flak. This probably had the effect of raising the level of violence on both sides; police confidence to behave forcefully and reactively was enhanced and from the demonstrators point of view, police protective equipment neutralised any inhibition on throwing things at them ... At **Lewisham** for example senior commanders had no idea of how to deploy shield equipped officers and ordered a wall of shields across the High Road.

A major way of reducing conflict was, this officer suggested, by forward planning and negotiation which, he held, had the potential to enable dissent and protest to be registered while avoiding direct conflict between the protesters and the police with the concomitant risk of compromising police legitimacy:

> The Miners' Union rally (held in London in 1992) had a very experienced Gold who had profited from debriefing the Poll Tax debacle. It was stressed during the briefing that officers should keep tight control of their teams and they were instructed to ignore the historical context - if the police had fought with the miners there would have been little public support or sympathy. The miners were almost totally supported by the public, conflict was to be avoided at all costs ... We must also recognise the police are a 'crowd' as well. It is important to stress to officers that the aggression is not at them personally, but at the uniform, at what it ultimately represents - they must recognise that they can still try to talk to people as individuals.

The importance of maintaining police legitimacy was also stressed by a Kent Inspector:

> As the operations are usually conducted within an environment that is essentially domestic, and there is a necessity not just to win, but to win in such a way that order can be restored and will eventually be largely self-sustaining, the police role must focus on recognising the potential for, and minimising the possibility of escalation. At the same time the police must not 'lose the streets', but must be in a position to go back in to preserve law and order. This is the importance of evidence gathering - order must be established first ... Evidence gathering can assist in de-escalation as officers get used to the idea that they do not have to react immediately to every crime occurring around them. This also has the advantage of encouraging officers to act with propriety as they will be videoed as well.

This particular respondent, however, raised another side to this particular issue by stressing that, in his view, legitimacy must, to some extent, be founded on public confidence in police effectiveness:

> The police can 'win by appearing to lose', as suggested by Sir Robert Mark (a former Metropolitan Police Commissioner), but this cannot go on indefinitely as the vast majority of the population will lose all confidence.

> Law-abiding society will get nervous if that ethic becomes totally pervasive and they may well demand a firmer line being taken that will reduce the possibility of normalising already difficult situations and localities. The police have to find solutions that pacifies and yet promotes sustainable confidence. No matter how high level the police response there is always the danger that dangerous disorderly behaviour will escalate to match it.

Such considerations are central to the problem of tactical escalation and de-escalation. The next respondent, a Kent Police Superintendent, provided some indication of how this balance might be achieved by way of due attention to both strategic and tactical training:

> Officers are trained to appreciate that there is a sliding scale of commitment to a riot situation. The issue is not so much about shield use as about training men to understand a demonstration and giving them a broad range of options. In doing this it is hoped that officers will not feel threatened and can make measured and sober judgements, instead of resorting to panicky overkill tactics. The more astute ground commanders in Kent are encouraged to make realistic casualty assessments - for example to wait until things are thrown before openly deploying shields. Table-top exercises are used in training to test when you raise technical intervention.
>
> The shield and numbered NATO helmet are the last option - there is little left after that - CS gas, baton rounds, firearms - hardly a desirable path to tread when ensuring public safety ... At **Tottenham**, at least, heavier weapons were kept on stand-by and not deployed despite the violence. That sort of restraint is necessary.

This officer gave an example of an operation with which he had been concerned and which, he felt, illustrated the value of co-ordinated strategic and tactical planning:

> A good example (of effective de-escalation and prevention of a potential public disorder situation) was a 'rave' pay-party that we dealt with. It was a seven-day intelligence operation and a one-day public order operation. Intelligence gathering prepared an appropriate level of response and effectively put the event off even before the public order policing was implemented.
>
> Because it became clear that the police were aware of the venue and the organisation and took action to put up notices at transport sites informing people that it was not going ahead, instead of 10,000 people turning up, only 35 car-loads did. The early intelligence operation led to the Assistant Chief Constable (Operations) (ACC) determining the budget for the enterprise, which limited the manpower available to police the event. This prevented over-policing and wasting resources.
>
> In this case a good deal of corporate responsibility was shared. The Silver commander had handled the intelligence gathering activities and was able to

> provide reasonably accurate information to the Gold ACC, as regards likely size of the event, and between them the policing cost was considered, costed and successfully implemented. Ideally **Gold, Silver and Bronze** should work like that - individual responsibility maintained within a corporate, supportive, structure.

A Kent Inspector with responsibility for strategic and forward planning also asserted the importance of social monitoring in order to sensitise officers to potential problems and make possible the planning of appropriate and measured responses:

> In Kent we have a whole range of issues that could lead to disorder. These are not merely large-scale urban crime disorders, but also smaller, perhaps unemployment related problems in the centres of the Medway towns. Our role is to predict and detect this potential as soon as possible in order to assist in the avoidance and containment of problems.

2. The present

All of our respondents were invited to describe and comment on current public order policing practice and make observations on what they saw as likely future developments in the field. Although the emphasis varied reflecting respondents' current professional responsibilities, most stressed the enormous strides that they saw as having been achieved in the fields of intelligence gathering and the development of multi-agency relationships, in negotiation and pre-planning, and in tension indication. Reference was also made to what were seen as concomitant improvements in command and control at strategic and tactical levels and in training and to the development of the technology for tactical deployment.

Negotiation strategies and developments

Most of our respondents agreed that significant developments had occurred in recent years in the raising of commanders' sensitivity to the importance of negotiating with figures and groups playing a leading part in the generation of crowds. This included communicating effectively with march and demonstration organisers and community leaders, whose opinion was likely to have an impact on the behaviour of a crowd both in terms of organisation and mood.

It was noted by an Assistant Chief Constable from the BTP that these aspects had always been important for his organisation because of its size, structure and responsibilities. For these reasons, he argued, his force has always been inclined to stress a positive relationship with the public and to be sensitive to the fact that it would almost invariably find its capacity to deal with disorder being stretched should this spread within the areas it policed. This respondent was proud that his force's policy of pre-planning and negotiation and of policing with the consent of those using public transport was in keeping with the emerging trend in public order policing in general as he saw it. He began by explaining how the BTP's style of policing had come about and how it was the cornerstone of its strategy for dealing with disorder:

> Policing the British Transport system involves the complexities of policing in both public and private environments. The force rarely polices events as such, but it does have responsibility for people moving to and from events, and for hundreds of thousands moving daily on the transport system for work, or recreation. The force is small and rarely has reserves on hand to handle large events unaided.
>
> Not having large-scale reserves, the force relies on interface with the public with very limited resources. Therefore the development of a special individual officer relationship with the public is a general trend and the facing of the public with paramilitary, unit style policing is rarely on the cards. The force works with this in mind under four broad strategies:
>
> i. rapport building by individual officers;
>
> ii. maintaining good intelligence about possible sites of trouble;
>
> iii. developing effective contingency plans and strategies;
>
> iv. maintaining good working relationships and communications with other interested agencies such as; other forces, the railway operators, event organisers (facilitating lawful assembly, demonstration, etc.), laying down the boundaries over which firm but fair policing will be applied and organising with the National Reporting Centre to bring in, or have home force officers on standby.

Stressing the complexity of the force remit and its relationship with other agencies, this officer continued by illustrating this with examples of the style of policing that he held to be desirable:

> 150,000 people work for London Underground and British Rail and the economic environment has made it likely that strikes will occur on these services. Relationships with trades union organisations therefore have to be kept at arms length, but mutual respect must be maintained. The British Transport Police have always successfully policed this environment and I do not think that we have ever prosecuted a railwayman. I and other force representatives meet with the Union executives two to three times per year and ASLEF (main railway union) diaries contain BTP phone numbers and details, as well as rights and regulations relating to the relationship of mutual respect between the organisations. We try to build a culture of genuine community policing.

> In cases of public disorder that has its roots outside, but spills into BTP territory, we liaise with home forces. For example, during the **Poll Tax** events and riot, the BTP kept mainline and underground stations clear and running and trouble free and liaised and gathered intelligence with the Metropolitan force ... (however) liaison was problematic on the street, although, thankfully, this did not have too great an impact on the BTP operation.

As regards the Metropolitan Police, a Chief Superintendent agreed that a great deal had changed in recent years with the development of multi-agency and 'partnership' approaches to the policing of public order. The reasons for this appear to be both awareness of the 'good practice' that such strategies constitute, and increasing attention to the personnel costs of policing public order in terms of training and the diversion of officers from other important aspects of policework at a time when emphasis is being laid on securing 'value for money':

> Over the years, there has been a commitment to develop a partnership approach. There is now much more emphasis on encouraging organisers to organise and steward effectively. The TUC (Trades Union Congress) are always happy to provide stewards and in most cases those taking part are increasingly prepared to take responsibility for ensuring safe and orderly events and demonstrations. The public appear to be happy to see the police more effectively deployed, but in smaller numbers and less obtrusively, as for example at football grounds. The partnership approach has always existed, but has had its ups and downs.
>
> The right to demonstrate is recognised and we are seeking more and more efficiency in policing, encouraging groups to provide stewards at critical points of marches and demonstrations. This leaves police free to intervene effectively when they are really needed. We tend to get much more support from organisers who are keen to keep their activities orderly in the interest of maintaining greater public support. People are now far more realistic. In the police service, we accept the limitations of our resources and are now increasingly aware of the necessity to liaise with other organisations. As a result we place a great deal of importance on contingency planning.
>
> There is much more support within the various communities than many police officers had previously thought and this has helped to avoid direct police/ public confrontation. Police officers are much better trained than before in 'heading off' potentially disorderly situations.

This respondent outlined the areas in which he considered the police had made considerable developments in recent years. We shall explore other opinions on these developments later in this chapter:

> There are a number of areas of recent development that are both a cause and result from this new awareness. Community liaison has made a great

> contribution to public order policing in terms of defusing the potential for spontaneous disorder in certain localities. The development of tension indicators are an important part of this relationship, the post-Scarman consultative committees have worked well in this context, both in terms of accommodating community police efforts and in maintaining potential sites of serious disorder. Sector policing also promises a good deal in terms of increased local responsiveness and accountability. The investigation of an event both pre- and post disorder is also important. Evidence gathering during riots minimises the necessity for contact and although it is not currently given a high priority, training at all levels and for all contingencies has improved markedly. This has led to better control of officers on the ground, helped by enhanced training, organisation and communications technology. The range and quality of equipment now available to us has also improved markedly.

Thus improvements in multi-agency work, negotiation and partnership are seen as parts of an overall development of the police capacity to deal with a whole range of disorderly and potentially disorderly circumstances. Our respondents stressed that efforts to negotiate multi-agency relationships and 'partnership' approaches were important for pre-planning events and for developing contingency plans and were a part of this wider picture of strategic development. Many forces, besides the Metropolitan Police, were developing institutionalised systems of forward planning and intelligence gathering so as to allow for effective and efficient deployment of resources. A major aspect of forward planning was the development of 'tension indicator' systems.

Monitoring tension for forward planning

The BTP Assistant Chief Constable pointed out, as mentioned earlier, that his force had maintained a positive working relationship with the trades unions in his area of operations, but also stressed how important it was that his force be aware of a host of other issues and circumstances that had a bearing on how and where he should deploy his officers. The BTP has a network of sources that feed information into its own information centre to facilitate the planning of deployment.

Other forces are also enhancing their forward planning potential. Although such planning can not predict spontaneous outbreaks of public disorder such as have occurred in recent years, an awareness of this potential, our respondents argued, could serve to shape general policing strategies as well as to facilitate contingency planning. An indication of the scope of this activity can be gauged from the following, from an Inspector charged with responsibility for forward planning:

> The Scarman Report on Public Disorder gave rise to a range of Home Office Circulars in 1982 and 1983 and this raised the issue that the police should look towards developing measures of community tension. This represents a major change in police sophistication and philosophy. Pre-Scarman we did

> not know what information gathering to predict disorder was. Criminal Investigation was relatively well developed and Home Office Circulars from the mid-eighties gave details of all kinds of festivals and feast days that could provide a focus for dissent and disorder and social conflict, but that was about all. Until comparatively recently, general police responsibility to be aware of political, economic and social change was underestimated.

A criticism that may be directed at the 'riot curve' described in Chapter I is that public order policing is in principle very much oriented towards returning an environment back to 'normality' with little consideration of the nature of that normality, or indeed the potential for that normality to generate future disorderly situations. Whilst an enhanced forward planning potential does not overcome this problem and the nature of local 'normality' remains a difficult issue, probably beyond the scope of the conventional policing remit, this officer described how deviation from local normality might be measured:

> A tension indicator system has now been developed in Kent and the results are regularly compiled and circulated around the force to raise local awareness among officers of all ranks. A major problem with tension indicators for public order purposes is that they are designed to measure deviations from 'normal' levels of disorder. We therefore need some indicator of 'normal levels of tension'. For example, if we have 4 officers assaulted, on average, per week in Gravesend and one week we have 7, does this represent a significant 'blip'? Perhaps - the quality of such information has to be judged by the local commander - we can only draw his attention to the data. We also look at local, national and international issues that may have local impact, or may impact on our mutual aid arrangements. The tension indicators used in Kent were developed here and are not necessarily the same as those used in other forces.
>
> One example was a sudden rise in unemployment in Kent. Over just a few days heavy concentrations of lay-offs were registered that needed attention. Network South East, for example, were laying off 1000, heavily concentrated in Kent. Blue Circle Cement were also making huge lay-offs. There were similar problems when the Kentish coal mines were closed - the jobs have gone, but the communities are still struggling on here.
>
> In addition, having the Port of Dover in the county means that we have to be aware of possible European influences, especially now regarding the channel tunnel, notably with regards to football crowds and the potential for international political protest (and more recently, of course, Animal Rights protests (Erlichman, 1995b).

The officer went on to indicate the range of issues covered by his system in Kent and described some of the methods used to monitor them:

> The methods that we use to raise tension-level awareness are many and varied, as are the sources from which we draw our information. We have an events

> notification procedure, established with regular 'customers'. For example we are notified by the military when there are to be military band performances. We also take notice of issues arising at certain times during General Elections and all of these notifications are subjected to threat assessment based on our other available intelligence. For example in the case of military performances we have in the past taken advice from Special Branch with regards to assessments of Irish Republican bombing activity.
>
> Similarly, when it was a current issue, we made ourselves aware of forthcoming meetings to discuss, or protest about the introduction of the Council (Poll) Tax. To do this we made contact with local councils and obtained assessments of community response. These forms of community consultation occur at a whole range of formal and informal levels. Hunts and hunt protest groups are also of interest if we are to allocate appropriate resources. We seek out information from both sides, both from formal groups and their organisers and through Criminal Investigation Department information. We have also to be aware of possible causes and sites of prison disorder in case we should be called in to assist the prison authorities. We have 11 prisons in Kent and three are remand prisons with potential for disorder that would require substantial commitment and accommodation.
>
> The international perspective can not be ignored, especially in this county. We have also been concerned in the past about conflict developing in India that may well impact on community relations in Britain. Further, for example, we may have to try and monitor the mood of French lorry drivers and their possible activities in Britain, which involves a high level of liaison, combined with grass-roots appreciation from international drivers, customs, etc., all with a view to European political changes that may impact on likely sites and forms of protest.
>
> Monitoring, liaison and consultation is a multi-layered exercise. We gather information right from the very top to grass roots level, and we are constantly trying to get our own staff to communicate back such intelligence and raise their awareness of such issues that they may not previously have regarded as being important or within their remit. Local community tension indicators are derived from all kinds of official liaison, but we do try to get individual local officers of all ranks to provide us with information to assist in forward planning. This is done by issuing both data collection and information dissemination circulars.

Stressing the importance of successfully disseminating the information gathered to other responsible officers, so as to maximise the sensitivity of the service to issues which impact on its tasks, this respondent continued:

> A vital element of a successful tension indicator system is the promotion of awareness among local officers of all ranks. This education to detect signs of possible disorder has been the subject of major effort by myself and those who work with me. We also have to be careful that the forward planning intelligence that we gather does not lead to us 'calling wolf' too often and hence leading to the whole system lacking credibility.

Whilst he had so far stressed the role of tension indicators in monitoring special events and circumstances, the officer was also keen to demonstrate the potential of such techniques for monitoring 'ordinary' police operations directed against crime and the possible impact of these operations on local communities to the point where they serve as precipitating factors, or 'flashpoints', for public disorder:

> Many drugs operations are notified to us for assessment of impact on community and possible tension development. I am also responsible for allocating operation names - to avoid possible sensitivity in labelling and to ensure local co-ordination. In the past twenty years training for public order and tension awareness has improved markedly, and although we still have a few officers with public order responsibility who train for only one day per year, this inexperience is declining. Their awareness of their diverse responsibility is also increasing. For example, at one stage we had warning of a migration of New Age Travellers into the area. We put officers on the ground to reassure the public that we were aware of the potential problem, even though the travellers did not eventually pose much of a problem (most did not turn up). They are also more aware of the fact that the police presence may be escalatory and make more sensitive deployment arrangements.[14]

A Kent Police Superintendent agreed with the forward-planning Inspector that, despite the information provided by the development of national centralised information-gathering and disseminating units, highly developed local knowledge was desirable, both when officers were actually deployed to deal with disorderly situations, and even more importantly in the tension-monitoring stage:

> You need people on the ground who are aware of how the national picture affects local issues and expectations. For example in Kent we could envisage some public order problems with the development of rubbish burning power stations on the patch. This would lead to demonstrations and marches and we must acknowledge people's rights to do these things. Therefore local knowledge and familiarity would be essential to open and maintain appropriate avenues of communication and negotiation. In such a case, we would talk to activists and group leaders to arrange marches and protests and, on this basis, could probably be confident that the vast majority would go along with our suggested plans, which would be a product of these negotiations.

---

14. ACPO organised a national intelligence system to monitor the movement and activities of New Age Travellers following a major impromptu festival held at Castlemorton Common, Worcestershire in June 1992. This is currently housed at NCIS.

> This also brings us to the issue of local knowledge in the development and interpretation of tension indicators. Sergeants and upwards should be brought into the tension indicator system - it would seem that an absolute quantifiable system is just impossible and a good deal of it should be based on 'gut feelings', but previously we have even been missing out on this type of information. We are now making strenuous efforts to gather all of this information and adequately interpret it. Under the current system, street officers are encouraged to feed back impressions of the 'temperature' of an area, i.e. how it differs from 'normal' and feed that information back to our intelligence officer who will in turn feed the information back up the system for prioritising and interpretation.
>
> Tension indicators take various forms and include:
>
> major crime;
>
> crime trends;
>
> missing persons;
>
> quality of life changes;
>
> racial incidents;
>
> political issues (local and national);
>
> major traffic changes;
>
> management information such as numbers of assaults on police and circumstances, complaints against the police, good policework and so on.

This respondent went on, crucially in our view, to draw attention to an aspect of the law which had a marked impact on forward planning potential:

> The provisions in the **Public Order Act (1986)**, making formal notice of crowd activity compulsory, is very useful in this field, especially in terms of forward planning for officer deployment. Of course those events at which trouble is to be expected are rarely notified, but notification of the others at least allows us to plan and leave a little in reserve for unexpected contingencies. Those that do notify can usually be accommodated and formal negotiated arrangements are generally adhered to by both the police and the demonstrators.
>
> There are problems with the Public Order Act (1986), in that it does not require that meetings in buildings be notified to the police so they do not automatically come into the tension indicator system. This exclusion can be

problematic. Folkestone Police at one time had information about a likely fascist presence in their area for a meeting. They gathered intelligence about it, but there was little that could be done to impose conditions legally because of the fact that the meeting was to be in a building.[15]

It is important for a local police commander to have his finger on the pulse of his community and to facilitate the venting of discontent by protest, and to make sure that it gets done in a manner that ensures the safety of marchers/ protesters, the rest of the public and the police.

Risk factor analysis is carried out by the Area Commander. This assessment, if serious, is then passed on to Gold (the senior officer responsible for the policing of a public order event, often for a large event the Assistant Chief Constable responsible for Territorial Operations), who has his hand on the allocation of budgets, etc. - his hand in these matters assists in maintaining a sense of proportion and discourages expensive, and possibly oppressive, over policing.

Organisation of command and control

In the course of the interviews our respondents gave their views on some of the changes that have been made in the police organisation of command and control in relation to public disorder. The problems that had led to dissatisfaction with early efforts at command and control were identified by a Kent Inspector:

> I was present in the central control room at the time of the Broadwater Farm riot. It was clear that something had gone badly wrong in terms of command and control judging by the reactions of those present. We were only there as observers and we were asked to leave.

A Metropolitan Police Chief Inspector outlined the rationale behind contemporary command and control organisation, while acknowledging that these arrangements had been in place at the time of events where they had not always worked as intended, in some cases even with disastrous results. What appeared to be the problem is that the '**Gold, Silver, Bronze**' command structure may be sound in theory, but does not work unless there is careful delineation and fulfilment of the responsibilities as laid down:

> Ideally the Gold, Silver, Bronze designation, should involve Gold being responsible for strategic planning; Silver controlling tactical deployments; and the Bronze commanders being those who direct officers who are actually engaged. 'On the day', Gold will usually be off site, in a control room making major strategic decisions, Silver may be 'on the ground', perhaps at an 'on

15. Generally, if this did not constitute a 'trespass', it would also seem to be excluded from the provisions of the **Criminal Justice and Public Order Act (1994)**. Some of the main provisions of this Act are outlined in the Glossary.

> site' command facility, and the Bronze commanders will be directing the officers 'on the ground'. This should avoid any conflict of command responsibility and direction, but in practice there have been immense problems relating to the interpretation of the system.

This line of argument was reinforced by a Metropolitan Police Chief Superintendent:

> The Gold, Silver, Bronze system has had its problems, but an experienced Gold is invaluable. So much depends on this person. Gold, Silver, Bronze works well provided the mixture of experience and personalities is right. The number of Bronzes is important as too many can lead to confusion among commanders on the ground. ACPO has recognised this difficulty. Gold is responsible for developing the strategy and should then let Silver 'get on with it', and only become involved by providing further strategic direction if the situation deteriorates in an manner which has not been envisaged and planned for.

The limited utility of the Gold, Silver, Bronze command structure to his force in view of its remit was explained by a BTP Assistant Chief Constable, although he did stress the necessity of his officers understanding and accommodating their activities to the requirements of local police commanders during major events:

> Gold, Silver and Bronze seem to be useful designations, but it is important to define precisely the expectations and location of these roles. BTP need to understand this, although, due to our remit, we are only required to exercise the Gold, Silver, Bronze management structure in force, on a very limited number of occasions. We do, however, have to understand how it may work in home forces in order to be in a position to respond and contribute to home department force operation, where the local Commander is designated 'Gold', but BTP input would either be at 'Silver' or 'Bronze' levels with appropriate designations.

Training

We were provided with a great deal of information by officers of Inspector rank and above who had responsibility for public order training. A Kent Inspector outlined the range of public order training courses run at the County and Regional training facility by the Kent Constabulary, ranging from the Silver and Bronze Commanders' courses and a Public Awareness course, to a Police Support Unit Commanders' course and Public Order Training for junior officers. Most other regions run various courses and the Metropolitan Police also run a number of training courses in public order policing at their 'Riot City' training unit in Hounslow, London. The Kent Inspector also informed us that there has been a shift to the training of highly specialised units alongside general training. This is particularly true of specialist evidence-gathering teams.

A British Transport Police Chief Inspector argued that there were problems with public order training. He felt that the setting of minimum standards and Mutual Aid requirements had tied the police to an inflexible set of tactics that could prove escalatory:

> A Gold senior officer would respond appropriately according to a four-fold categorisation of forms of dissent (as mentioned in Chapter II, these would be political, industrial, festival and urban). The categories would be policed differently according to the social context and so on, but if problems do occur it seems unfortunate that the senior officers only have one basket of responses to fall back on.

Technology

We now turn to the crowd and riot control technology that is deployed by the police in England and Wales, taking a Kent Inspector's description of the No. 5 Region and Kent experience as typical. The basic public order police unit used in actual and potential disorders is the **Police Support Unit** (PSU), which may also be deployed in Mutual Aid situations. The structure of PSUs has changed recently to allow for closer control by supervising officers, as described here by a Kent Police Inspector:

> Since 1991 a PSU has 22 officers. The previous arrangement was an Inspector, 2 sergeants and 20 constables. Now, there is an Inspector, 3 sergeants and 18 officers, improving the level of supervision. This is technically a national arrangement, but there may be some variations. Mutual aid arrangements do not cover dogs, horses, CS (gas), baton rounds, long batons or arrest teams - these are tactical options provided by specialist units.

a) Batons

The Inspector went on to say that:

> In Kent there is training in the use of long batons, introduced in 1985. This is provided for Territorial Support Group personnel. In the old days some forces issued night sticks with little or no training, but in 1985 this was formalised in a Home Office circular which specified the use of long batons and training. There are designated levels of authority to authorise the use of long batons, as well as other specialist equipment. In Kent PSU commanders are trained in long baton authorisation and are designated officers who can authorise their use. The side-handled baton, however, is according to the US experience, of limited use for public order work; it is more a weapon for personal protection.[16]

16. There are now a variety of side-handled, expandable and longer batons in routine deployment by the different police forces in England and Wales, replacing the traditional 15½" wooden truncheons and introduced in response to calls from ACPO and the Police Federation.

b) Baton rounds

These are plastic bullets (although there have in the past been wooden and rubber varieties), about 11cm in length and 4cm in diameter (Sadler, 1992:112). Plastic bullets were first introduced in Northern Ireland in 1972, and by 1985 had already killed 12 people, 6 of whom were children (BSSRS, 1985:72). They were deployed but not used in England in 1985 at the Broadwater Farm estate riot.

In contrast to most other European countries, firearms are not routinely carried by police in England and Wales. There is an increasing trend in this direction, however, with Armed Response Vehicles being deployed nation-wide in recent years and guns being issued to officers more often in special circumstances. We should note, however, that a recent survey of Police Federation members came out overwhelmingly against their routine deployment (D. Campbell, 1995).

c) Gas and sprays

Although used by the British in Cyprus in 1958 and later in Northern Ireland, CS gas (orthochlorobenzylidene malonitrile, and named after B.B. Corson and R.W. Stoughton, who invented and developed it in 1928 - Ackroyd *et al.*, 1977:214-215) was not used in England until the Toxteth riots of 1981. It was deployed but not used at Broadwater Farm, Tottenham in 1985. There have more recently been trials with disabling gas sprays which could be routinely deployed. One of these, 'pepper' (capsicum) spray, has not been adopted following trials, it having been found to be potentially harmful for recipients suffering from asthma and those who are pregnant (Travis, 1995). CS gas in spray form will shortly undergo trials 'on the street' with 18 police forces (Worsnip, 1995).

d) Shields

As we have noted, the deployment of shields was seen as a watershed in the development of public order policing in England and Wales. They have continued to be developed as public order strategy and tactics have shifted towards greater mobility. A Kent Inspector with responsibility for public order training described the development of shields:

> There are many developments underway in shield technology ... The 'Kent shield' (Armadillo) has corrugated edges that can link together. Round shields may be used by arrest teams as a part of the basic five-man unit concept - generally three would have long shields with two behind with short ones. In contrast, long shields have their strengths in terms of physical protection, but tend to be tactically cumbersome and unwieldy.
>
> Two more shields are, firstly, a smaller one which is not so much a public order shield, but is carried in vehicles for immediate protection - for example,

> keeping an 'angry man' at bay. They are not used by the whole force, but are a stop-gap that can be carried in a vehicle and while it is being deployed it is envisaged that more appropriate resources will be called up. They really offer limited, basic protection and have little containment capacity for formal public order work. Secondly, there is the Kent pattern linking shields that can attach to the Kent shield vertically. Both of these are alternatives to the French *gallet* ovoid shield.

Given the current structure and funding arrangements of the police in England and Wales, the technology deployed in each area varies in accordance with perceptions of local need and costs.

e) Protective clothing

The stress on mobility and the deployment of shorter shields has placed renewed emphasis on supplementary forms of protection. Gone are the days when there was a rush to obtain shin, groin and knee protectors from sports stockists following the 1976 Notting Hill Carnival. In addition to such items, PSUs are issued with flame-retardant clothing and NATO-style helmets and visors. More recently, there has also been a routine issue of either stab or bullet-resistant vests.

f) Evidence gathering

The risk that the appearance of riot-equipped officers may well prove to be escalatory is by now well appreciated, as indeed is the idea that however well equipped they are it may be difficult to contain a situation rapidly. This awareness as well as its use in subsequent identification and prosecution has led to an increasing concentration on evidence gathering. A Superintendent in the Metropolitan Police provided an insight into current thinking in this field:

> Developments in the field of on-site evidence gathering reduce the immediate imperative and make court cases more tenable. Officers committed to order maintenance have limited ability to obtain and hold evidence for arrest. Information cells and gatherers help to back up officers on the street and have time and facilities to retain and gather evidence. For example, when I was operational at the **Poll Tax** riot, of the many, many offences that I witnessed, I only had clear memory of two from which I could confidently have made a statement.

However, a Detective Superintendent in the Metropolitan Police, in an article describing and promoting the role of senior investigating officers at public disorder events, recently acknowledged that:

> Frankly, the police record for successfully prosecuting those who engage in serious public disorder is dismal ... the police service can, and should respond to

> changes in evidential and investigative standards, be awake to technological developments and be prepared to properly train those on whose shoulders they place the onerous task of public disorder investigation (Williams, 1994:134-5).

g) Communications

The changes in technology during the period on which this book focuses have led to communications between commanders and officers on the ground becoming an important issue. This extends also to the establishment of common minimum standards of communication to facilitate co-ordinated action between police services - a trend which goes against the traditional emphasis on local police autonomy of policing in England and Wales but which is, through gradual change and pragmatic adjustments, becoming an increasing feature of strategic police work. The current situation, as regards both aspects of communication, was outlined by a Kent Inspector:

> As clearer understanding of roles is improving, so too are technical aids to facilitate local control. PSU helmets may, for example, now be equipped with a bone microphone for unit commanders to broadcast to PSU officers.
>
> Common minimum standards for mutual aid provision are still rather varied, but communication between forces has improved since its introduction and future possible regional, or national reorganisation may well improve this.

3. The future

Opinion concerning future trends varied from a Metropolitan Police Chief Superintendent's view that forces were 'about there', bar some organisational and training aspects that required sharpening, to a Kent Inspector's view that something approaching a 'third force' along European lines would gradually emerge.

The Metropolitan Police officer commented:

> Training is much better than it used to be and we are now more equipped to do the job. We still need to develop and implement techniques which will reduce the potential for public disorder. Safety as an issue, although first recognised at Red Lion Square (in 1974) was highlighted by the Hillsborough (stadium) disaster (in 1989). As a result the accountability of police, and more specifically the personal liability of chief officers, will ensure this area of policy is a priority. The public appear to want police officers on the beat rather than policing demonstrations. If we follow this trend, it will hopefully lead to a further decline in macho image of the service and a change in culture.

A Kent Inspector, with responsibility for public order training, generally agreed, although from a somewhat different perspective:

> A Third Force development currently appears to be unlikely. The only real circumstances in which this is likely is if reorganisation takes a truly national character. Some developments are promoting even greater levels of co-operation. The Channel Tunnel is one example ... The Tunnel, however, is promoting liaison with the French, notably over firearms use, football disorder and the possibility of French farmer protests. Training for alternative policing methods in public order, such as evidence gathering is also on the increase.
>
> It would appear that events such as Brixton and Broadwater Farm are on the decline and the Trafalgar Square Poll Tax incident was perhaps rather exceptional. Currently incidents seem to be more localised and the demonstrators/rioters more likely to be fewer in numbers and the disorder more sporadic. At Blackbird Leys, Coventry and Bristol a few hundred were in the crowd, and that was problematic. In such circumstances the old five-man slow response was inadequate; they are now more mobile. Mobility is the key and this has developed, and is developing, in an incremental way over the years.

Another Kent Inspector, with local public order responsibility, provided a further view of future developments:

> In some senses I would argue for a 'third force', specially trained and very carefully selected so that we can avoid the excesses of some countries close by. Such a force could provide a swift, effective response to many major incidents, not just those posing problems for the maintenance of public order. Self control must be a major selection point as well as physical, and to some extent intellectual abilities. An awful lot would be expected of such a force in the British experience. They would have to make very quick, but also sensitive judgements and be constantly aware that their purpose is to operate, then to leave the situation in such a way that it can be picked up again by the local police. A major drawback is the expense of such a unit and what to do with it when it is not operational. Its over-use must also be discouraged, in case local officers become too ready to abandon their efforts and deploy it, bringing in the problems of escalation and reducing the possibility of restoring self-sustaining order in the short to medium term.

These comments provide a 'snapshot' of police perspectives on public order policing in England and Wales and illustrate the directions of current thinking and practice in the field of strategic and tactical planning and deployment. In the conclusion we draw these observations together, stressing the need to understand the 'interactive' nature of the public order policing function. We suggest that an understanding of the dynamic relationship between public order policing and social change is necessary if public order policing is to remain sustainable and as legitimate as possible in a society exhibiting a complex fragmentation of interests and perspectives.

Conclusion

In this work we have examined the changes that have occurred in the area of public order policing in England and Wales in three main strands. In Chapter I we reviewed a range of existing and possible theories of the nature of crowds that might underpin a coherent public order policing strategy, a strategy that needs to embody the concept of securing consensus in support of police activity, in which the idea of exercising minimal force plays an important part. We have argued that a credible theory of crowd behaviour must accept the essential rationality of most crowd participants. Further, the dynamic potential of crowds must be acknowledged and the possible impact of police and other action on this dynamic process appreciated in order to arrive at a workable solution to the problem of controlling public disorder. Chapter II drew attention to police responses and their contribution to events, as well as to factors such as the forms of political confrontation that lead to the development of the crowd and shape the antecedent circumstances, including prior policing experience. The product of our interviews with a range of serving police officers with diverse public order policing responsibilities is presented in Chapter III. Officers commented on recent developments in public order policing, on the technological advances that accompanied and contributed to these developments and on the changing training arrangements, as well as the introduction of more covert, but proactive, intelligence gathering activities and forward planning. The interviews go some way to illustrating the diversity of opinion over these issues, although they largely confirm, albeit to different degrees and with different emphases, the pattern of change that we outlined in Chapter II as a result of our historical analysis.

We will conclude by drawing together some of the main threads of this work, especially where they indicate directions of change. There is little doubt that the method of policing public order situations has changed dramatically over the last 25 years. What we have attempted to indicate here is not only the way in which it has changed, in terms of strategy and tactics, but also to locate that change within some sort of dynamic process. This process, we suggest, involves amongst other things a relationship between the police and the 'crowd' which has a potential for escalation and de-escalation and is affected to a major extent by the method of policing. As we have argued in Chapter I, the crowd should not simply be regarded as an unthinking and irrational mass. It does have a rational and purposive dimension in action, and policing training needs to take this into account. However, we have suggested also that in order to understand any given instance of disorder fully, a number of factors need to be taken into consideration, be they Scarman's 'antecedent conditions', Waddington's six levels of tension contributing to the emergence of a 'flashpoint' or Smelser's 'structural conduciveness and strains'. The 1980s and 1990s have seen outbreaks of violence in some of the more socially deprived areas in England and Wales. Some of these sites of disorder are inner-cities, others are depressed areas on the outskirts of major cities that have felt the chill wind of economic recession and the full impact of social and political marginalisation, often in terms of increased crime, victimisation and police attention. In examining our chosen instances of disorder within an (as we shall explain, not necessarily all-encompassing) four-fold typology, we have attempted to indicate some of these wider contextual matters without seeking to go into great depth, but limiting ourselves to an empirical 'snapshot' review.

We have been rather more concerned with looking at changing strategy and tactics in terms of certain watersheds or conjunctures, suggesting that these have generally been police responses to actual or perceived changes in the form of public dissent. The problem here, of course, is the mutually causative relationship between these two factors, which, together with the inevitable impact of technological determinism can easily give rise to an escalatory 'spiral of violence' (Wright, 1978). However, it is important to stress that the trend of strategy and tactics for policing public order has not simply been towards the use of a harder fist. There is additionally a move within public order policing towards pre-emptive and proactive investigation and prediction, whilst at the same time developing a more flexible (in terms of different levels of response), highly trained and mobile public order force.

This two-pronged approach, we would suggest, is not only due to an awareness on the part of police policy-makers of the problem of escalation (although from some of our instances in Chapter II we would argue that by and large this does exist), but also to the limits on resources. On the one hand, it is more economical, in respect of deployment of resources, if for example the various types of potential disorder are policed according to the level(s) of threat they actually pose; just as effectively matching the level of potential threat with that of control would seem to contain the situation rather than cause it to escalate. On the other hand, it is more economical if one can predict when 'spontaneous' disorder is likely to occur in urban areas, for example; such forecasting could lead to greater use of contingency plans for accommodation, or rapid containment (or even effective deployment of evidence-gathering teams for arrest after the event) which should help to maintain public confidence in the police.

Changes in public order policing strategy and tactics are also affected by, and may even play a part in determining, the changing organisation of the policing system. In England and Wales there is an increasing tendency towards, on the one hand, centralisation and regionalisation (Johnston, 1993; King, 1995; Loveday, 1992) and, on the other hand, more decentralised community policing, through structures such as sector policing for example. In the long term, such reorganisation may well lead to the development of a 'third force' at one level, and a more pervasive network of covert surveillance, intelligence gathering, information storage, 'sensitive' categorisation, tension indicators, targeting and, in effect, 'criminalisation' at the other (P. Cohen,1979; S. Cohen, 1985; G. Marx, 1988; Stenning, 1989). On this basis, the intrusive aspect of the velvet glove could have escalatory tendencies also. Intensive detective work, surveillance and information gathering may affect communities and groups in such a way as to add to the impact on them of other unfavourable circumstances and the experience of economic, political and social marginalisation that serve to damage the image of the police, and compromise efforts to secure policing by consent through genuine community involvement and co-operation. The strategic relationship between proactive community initiatives, reactive and preventive crime control and public order policing has to be constantly borne in mind and reviewed if policing is to carried out legitimately, by public consent and with the exercise of minimal force.

Whilst it is acknowledged that co-operation and negotiation by the police and the use of stewarding by organised groups of protesters have done much to defuse conflict, attention needs to be drawn to the debate on contemporary social change that appears likely to complicate and could compromise existing arrangements, and at the very least bring into question the four-fold typology of disorder referred to earlier (or five-fold if we were to include sporting events). This debate is focused on change within something broadly and rather contentiously termed 'postmodern' society.

The main thrust of most 'postmodern' discourse[17] has been that the ideas and social and institutional structures on which western, and many western-influenced, societies have been based for the past three hundred years may now be being seriously questioned. The founding ideas of the 'modern' period are generally accepted as having been developed during and since the period of the European Enlightenment in the late eighteenth century and are typified by an adherence to a broadly, yet philosophically limited, 'scientific' rationality embedded in social and political institutions. The assertion of diverse human interests and experiences, beyond those raised by the Enlightenment such as social class interests as defined by an individual's or group's position in society, has taken place alongside varied efforts to make sense of the world from increasingly self-aware perspectives. Issues of race and gender are but two examples which have come to prominence over the past thirty years or so, but increased intellectual and emotional ferment and exploration have given rise to a new assertiveness among groups exploring humanity's relationship to the material world as manifested in the 'green', environmental, anti-nuclear and animal rights movements.

The same might also be said of those who choose to pursue alternative lifestyles such as 'New Age Travellers'. In this case the choice may be due either to a commitment to an alternative view of a fulfilling way of life, or to the fact that an individual's material choices are circumscribed by a lack of conventional opportunities as a result of crises in capitalist economies, and to whom a 'New Age' lifestyle seems preferable to unemployment, underemployment and social marginalisation within one geographical area. All of these interests have a long pedigree and have previously been accommodated, marginalised or ignored by the prevailing structures of modern society which has

17. Important critical contributions to this debate, both for and against the interpretation of contemporary western society as 'postmodern', are contained in Callinicos (1989); Crook *et al.* (1992); Nelken (1994) and O'Neill (1995).

generally served, at least partially, to legitimate action taken by nation-states and their control agencies, such as the police, to suppress their demands and protests in the interests of an all-encompassing view of modern rationality.

In our view the changing form of dissent, whether or not this actually constitutes a 'postmodern' condition, appears likely to pose a succession of problems for a public order policing policy which attempts to adjudicate between interests and to moderate the impact of dissent by reference to some view of a manageable social consensus, and which legitimates the police 'taking sides' to protect one, generally established, interest over another. In the contemporary conditions, the potential for clashes of interest and perspective are enhanced by an emerging social and intellectual diversity and it is likely that public order policing will be a spectacular focus of this ferment.

As mentioned earlier, awareness of this leads us to question the four-fold categorisation of public order events typifying the development of the public order policing 'response' in England and Wales. Our interviewees' general acceptance of the validity of this typology suggests that, despite significant changes and efforts to accommodate shifting patterns of social and political legitimacy, and to adopt new and more subtle forms of control, the police are still engaged in dealing with political and social circumstances which have perhaps to some extent been overtaken. The emergence of seemingly new social and economic trends provide serious challenges to the maintenance of a broad public acceptance of the legitimacy of police action.

It would appear, then, that the contemporary condition provides serious new challenges to the law and the manner in which it is implemented and enforced. For many writers, notably those of a Marxist or critical orientation, this has always been problematic. Whose interests are represented by the law and by the manner of its enforcement has always been a moot point, but police organisations in England and Wales have generally been able to rely on the mobilisation of some form of consensus to contain an overt expression of one, generally the socially and politically more powerful, group over the others. In England and Wales this legitimation has generally been achieved by reference to concepts such as 'the rule of law' or to some vague, but pragmatically supportable 'popular will', reinforced by established political groups and the routine representation by the mass media, with the role of the police being somehow depicted as 'neutral arbiters' of law and order.

In a society that has seen itself broadly divided by class interests, the emerging challenge of the contemporary situation makes it a great deal more difficult for those in authority to legitimate their power by reference to the demands of a 'general will'. The growing fragmentation of economic, social, political and philosophical interests challenges the representative pretensions of established political institutions and severely weakens the legitimacy any efforts to assert that there is a general will emanating from them. A variety of political, moral and philosophical perspectives have emerged and gained varying degrees of acceptability and credibility for accommodating or confronting this apparent crisis.

As we have noted, there have always been tensions in, and resistance to, the dictates of rational modernity and these can be seen to have manifested themselves throughout modern history, as we briefly noted in Chapter I. This discontent has been in evidence even during the period that we have covered in this book. The anti-nuclear protests of the early 1960s, the anti-Vietnam war demonstrations and the manifestation of the problems facing black youth in disorder, as well as conflict in the field of industrial relations as economic pressures threaten to transform the nature of workplaces and their associated local communities, have all been identified as 'watershed' developments which have been taken on board by police planners. In England and Wales, however, an even broader range of discontents within modern society can be identified

Even without reference to the evidence of an enduring modern 'history of respectable fears' described by Pearson (1983), we might witness the public disorder manifested following changes in the youth labour market in the early 1960s focusing on the emergence of youth subcultures whose activities and participants were successfully 'demonised', arguably resulting in something of an over-reaction by the mass media, politicians and ultimately the police and generating a 'moral panic' (Cohen, 1980). A similar tale may be told of black youth, notably during the 1970s as documented by Hall *et al.* (1978), Pryce (1979) and more recently by Keith (1993). The pursuit of alternative lifestyles and interests has also more recently attracted police attention as a possible public order problem, as reflected in the Criminal Justice and Public Order Act (1994) which criminalises trespass on private land.

This effort to pursue 'alternative lifestyles' has continued since the 1960s despite regular conflict with the police and established local interests. Such conflict has had varied sites, ranging from the 1974 Windsor Festival to Stonehenge. The 'Battle of the Beanfield', which resulted in 500 arrests and significant damage to the property of New Age Travellers as police closed the Twelfth Stonehenge Festival in 1985, may be regarded as another 'watershed'. All criminal charges arising from the Battle of the Beanfield were dropped following proceedings lasting until 1987, but Section 39 of the Public Order Act (1986) gave the police enhanced powers to deal with New Age Travellers, as has the Criminal Justice and Public Order Act (1994) mentioned below.

In contrast to the manner in which a number of confrontations involving New Age Travellers in England and Wales developed, it would seem that West Mercia police adopted a distinctly 'softly softly' policy at Castlemorton Common in June 1992. Many of the participants at this event had migrated from the site of the Avon Free Festival which had been prevented from occurring as a result of the activity of Avon and Somerset police. The West Mercia operation reflected changes in public order policing in that covert methods of detection were used and surveillance and intelligence-gathering resulted in officers being informed of likely developments and their likely impact. Some effort was also made at 'community policing' in order to negotiate a peaceable outcome to the event and to inform anxious local residents of the rationale behind police action. Although many local interests felt aggrieved by the police handling of Castlemorton Common, it is clear that the major potential for public disorder was averted. Those who were arrested for 'conspiracy to cause a public nuisance' were acquitted following court hearings in 1994 (Judge, 1994).

Despite the watershed developments in the strategic and tactical management of public order policing between the 'Battle of the Beanfield' and Castlemorton Common, local practice and outcomes vary as contemporary lifestyles and established interests conflict. The Criminal Justice and Public Order Act (1994) has been used to tighten up the provisions of the Public Order Act (1986). In the immediate context it would appear that this legislation was developed in order to provide the police with more powers to take immediate action against New Age Travellers, participants at 'raves', environmental protesters aiming to dispute construction and development projects, animal rights activists and those aiming to sabotage 'field sports'. Whilst these new powers constitute strategic and tactical tools to assist the police in curtailing activities which disturb and offend generally established interests, it is emerging that in a developing society the interests against which they are designed to operate are less and less seen as socially and politically marginal. Indeed, the legislation was itself the subject of a broad range of protests, most notably resulting in major disorder following a rally in Hyde Park in October 1994.

Whilst more sensitive public order policing, involving intelligence gathering and negotiation may be a pragmatically sound intermediate strategy, we wonder how otherwise 'respectable' demonstrators will view the prospect of being the subject of police surveillance and by implication 'suspect'. Might this not lead to a widespread awareness of a more covert form of control and threaten the hitherto accepted, if traditionally class-biased, basis for police legitimacy? If political and economic realities lead to an increasing number of people who may have regarded themselves as belonging to the 'respectable' elements in society being forced either to take on the traditional image of the 'rough', or to adopt 'alternative' lifestyles, where is the basis of legitimacy for contemporary policing? As Reiner concludes when writing of the implications of the development of postmodern conditions for policing in general:

> Post-modern culture may have eclipsed the Enlightenment's modern conceptions of social justice, as well as the more ancient prophetic religious ideas which modernism had displaced earlier. But certain harsh realities will not be pushed aside. As Los Angeles, the modern world's dream factory, showed us in May 1992, the backlash of the oppressed can turn complacent reveries into nightmare. To paraphrase Rosa Luxemburg, in the final analysis the only alternatives are social justice or barbarism. Unfortunately at present, the odds seem strongly to favour barbarism (Reiner, 1992b:781).

The sight of apparently genteel and ineffectual elderly people being 'manhandled' by police officers at pickets in Shoreham-by-Sea, West Sussex and Brightlingsea, Essex seeking to disrupt the exportation of sheep for slaughter and live calves for the veal trade (Erlichman, 1995a); 3,000 youths fighting 'a pitched battle with police' after the latters' intervention to stop a 'rave' party in Luton, Bedfordshire (Pendlebury, 1993); serious police and demonstrator clashes following a march and rally in Hyde Park, London by over 50,000 people to protest against the Criminal Justice Bill (Campbell and Travis, 1994); the distribution of a letter from the Assistant Chief Constable of Essex to all households in Brightlingsea threatening prosecution under the Public Order Act (1986) 'to restrain unlawful activity ... (such as) obstruction by walk(ing) in front of vehicles (containing sheep and veal calves) on the highway' (Bellos, 1995); the eviction under section 61 of the Criminal Justice and Public Order Act (1994) of 100 'travellers' from 'one of the most established' sites (set up 8 years ago) in southern England, Semley Woods in Wiltshire (Chaudhary, 1995) and the death of a woman, Jill Phipps, in an accident resulting from protesters' efforts to prevent a lorry delivering veal calves to Coventry airport in early 1995 (Hornsby, 1995) are all incidents which demonstrate the increasing complexity of public order policing in contemporary England and Wales.

Whilst the police in England and Wales have developed considerable sophistication in dealing with public order, one questions whether the innovations in community consultation (in the broadest sense of 'community'), surveillance, command and control and technology will lead to the 'lid being kept on'; or will the complexity of the task, the pressure on scarce resources, the fragmentation of interests and the erosion of the base for popular acceptance of the legitimacy of police action simply overwhelm the structures that have been developed and barbarism result; or, and it is this we think more likely, will, given the dynamic relationship between policing and dissent, new forms of both develop?

Bibliography

ACAB (1990) *Poll Tax Riot: 10 hours that shook Trafalgar Square,* London: ACAB Press.

Ackroyd, C., Margolis, K., Rosenhead, J. and Shallice, T. (1977) *The Technology of Political Control*, Harmondsworth: Penguin.

Allport, F.H. (1924) *Social Psychology*, Boston: Houghton Mifflin.

Amey, G. (1979) *City Under Fire: the Bristol riots and aftermath,* Guildford: Littleworth Press.

Beckett, I. (1992) 'Conflict management and the police: a policing strategy for public order', in T. Marshall (ed.) *Community Disorders and Policing: conflict management in action,* London: Whiting and Birch: 129-139.

Bellos, A. (1995) 'Police clamp on animal demos', *The Guardian*, 10 April.

Benyon, J. (ed.) (1987) *Scarman and After: essays reflecting on Lord Scarman's report, the riots and their aftermath,* Oxford: Pergamon Press.

Benyon, J. (1993) *Disadvantage, Politics and Disorder: social disintegration and conflict in contemporary Britain*, Occasional Paper in Crime, Order and Policing, Leicester: CSPO.

Benyon, J. (1994) *Law and Order Review 1993: an audit of crime, policing and criminal justice issues*, Leicester: CSPO.

Benyon, J. and Solomos, J. (1987a) 'British urban unrest in the 1980s', in J. Benyon and J. Solomos (eds) *The Roots of Urban Unrest,* Oxford: Pergamon Press: 3-21.

Benyon, J. and Solomos, J. (eds) (1987b) *The Roots of Urban Unrest,* Oxford: Pergamon Press.

Borrell, C. (1976) 'Police Commissioner defends use of 1,598 men at Notting Hill', *The Times*, 1 September.

Bowden, T. (1978a) 'The police response to crisis politics in Europe', *British Journal of Law and Society*, 5 (1): 69-88.

Bowden, T. (1978b) *Beyond the Limits of the Law*, Harmondsworth: Penguin.

Brake, M. and Hale, C. (1992) *Public Order and Private Lives: the politics of law and order,* London: Routledge.

Brearley, N. (1991) 'Riot control - understanding crowd psychology', *Intersec,* 1 (6).

Brearley, N. (1992) 'Public order, safety and crowd control', *Intersec,* 2 (1).

Brewer, J.D., Guelke, A., Hume, I., Moxon-Browne, E. and Wilford, R. (1988) *The Police, Public Order and the State: policing in Great Britain, Northern Ireland, the Irish Republic, the USA, South Africa and China,* Basingstoke: Macmillan.

Brightmore, C. (1992) *Urban Rioting in Latter Day Britain: an analysis of causes, tension monitoring, disorder prediction and prevention* (unpublished MA dissertation), Leicester: University of Leicester, CSPO.

Broadwater Farm Inquiry (1986) *Report of the Independent Inquiry into Disturbances of October 1985 at the Broadwater Farm estate, Tottenham,* London: Karia Press.

Brogden, M., Jefferson, T. and Walklate, S. (1988) *Introducing Police Work,* London: Unwin Hyman.

Brogden, M. and Shearing, C. (1993) *Policing for a New South Africa*, London: Routledge.

Brown, J. (1982) *Policing by Multi-Racial Consent: the Handsworth experience,* London: Bedford Square Press/NCVO.

BSSRS (British Society for Social Responsibility in Science) Technology of Political Control Group (1985) *Techno Cop: new police technologies*, London: Free Association Books.

Bunting, M. (1991) 'Violence driven by car thefts', *The Guardian*, 3 September.

Bunyan, N. (1992) 'Armed police hunt Salford estate gunman', *The Daily Telegraph*, 8 July.

Bunyan, T. (1976) *The History and Practice of the Political Police in Britain*, London: Quartet.

Bunyan, T. (1981) 'The police against the people', *Race & Class*, XXIII (2/3): 153-170.

Callinicos, A. (1989) *Against Postmodernism: a Marxist critique*, Cambridge: Polity.

Callinicos, A. and Simons, M. (1985) *The Great Strike: the miners' strike of 1984-85 and its lessons,* London: Socialist Worker.

Campbell, B. (1993) *Goliath: Britain's dangerous places*, London: Methuen.

Campbell, D. (1993) 'Condon amends anti-race demo', *The Guardian*, 15 October.

Campbell, D. (1995) 'Police balk at guns as routine on the beat', *The Guardian*, 16 May.

Campbell, D. and Travis, A. (1994) 'Police blame anarchists as marchers call for inquiry', *The Guardian*, 11 October.

Carvel, J. (1991) 'Police chief rejects hardship link', *The Guardian*, 21 September.

Cashmore, E. and McLaughlin, E. (eds) (1991) *Out of Order? policing black people*, London: Routledge.

Chaudhary, V. (1995) 'Police order eviction of site', *The Guardian*, 8 April.

Chibnall, S. (1977) *Law and Order News*, London: Tavistock.

Christian, L. (1983) *Policing by Coercion: the Police and Criminal Evidence Bill*, London: Greater London Council, GLC Police Committee Support Unit.

Clutterbuck, R. (1978) *Britain in Agony: the growth of political violence*, London: Faber and Faber.

Cohen, P. (1979) 'Policing the working class city', in B. Fine, R. Kinsey, J. Lea, S. Picciotto and J. Young (eds), *Capitalism and the Rule of Law: from deviancy theory to Marxism*, London: Hutchinson: 118-136.

Cohen, S. (1980) *Folk Devils and Moral Panics*, London: MacGibbon and Kee.

Cohen, S. (1985) *Visions of Social Control: crime, punishment and classification*, Cambridge: Polity Press.

Colman, A. (1991a) 'Psychological evidence in South African murder trials', *The Psychologist*, 4: 482-486.

Colman, A. (1991b) 'Crowd psychology in South African murder trials', *American Psychologist*, 46: 1071-1079.

Cooper, P. (1985) 'Competing explanations of the Merseyside riots of 1981', *British Journal of Criminology*, 25 (1): 60-9.

Coulter, J., Miller, S. and Walker, M. (1984) *A State of Siege: politics and policing in the coalfields,* London: Canary Press.

Cowell, D., Jones, T. and Young, J. (eds) (1982) *Policing the Riots,* London: Junction Books.

Craig, Y. (1992) 'Policing the poor: confrontation or conciliation?', in T. Marshall (ed.) *Community Disorders and Policing: conflict management in action,* London: Whiting and Birch: 77-86.

Crick, M. (1985) *Scargill and the Miners,* Harmondsworth: Penguin Books.

Criminal Justice and Public Order Act (1994), Chapter 33, London: HMSO.

Crook, S., Pakulski, J. and Walters, M. (1992) *Postmodernization: change in advanced society*, London: Sage.

Cumberbatch, G., McGregor, R., Brown, J. and Morrison, D. (1986) *Television and the Miners' Strike,* London: Broadcasting Research Unit.

Dear, G.J. (1985) *Handsworth/Lozells, September 1985. Report of the Chief Constable West Midlands Police to the Secretary of State for the Home Department,* Birmingham: West Midlands Police.

Department of Employment (1980) *Code of Practice on Picketing,* London: HMSO.

Diener, E. (1979) 'Deindividuation, self awareness and disinhibition', *Journal of Personality and Social Psychology,* 37: 1160-1171.

Diener, E. (1980) 'Deinviduation: the absence of self awareness and self regulation in group members', in P. Paulus (ed.) *The Psychology of Group Influence,* Hillsdale, NJ: Earlbaum.

Dipboye, R. L. (1977) 'Alternative approaches to deindividuation', *Psychological Bulletin,* 84: 1057-1075.

Donegan, L. (1995) 'Muslim leaders warn of other cities on the verge of violence as police give up', *The Guardian,* 12 June.

Donelan, R. (ed.) (1982) *The Maintenance of Order in Society: a symposium*, Ottawa: Canadian Police College.

Dromey, J. and Taylor, G. (1978) *Grunwick: the workers' story,* London: Lawrence and Wishart.

Dummett, M. (1980) *The Death of Blair Peach: supplementary report of the unofficial Committee of Enquiry,* London: National Council for Civil Liberties.

Dunning, E., Murphy, P., Newburn, T. and Waddington, I. (1987) 'Violent disorders in twentieth-century Britain', in G. Gaskell and R. Benewick (eds) *The Crowd in Contemporary Britain,* London: Sage: 19-75.

Emsley, C. (1991) *The English Police: a political and social history,* Hemel Hempstead: Harvester Wheatsheaf.

Erlichman, J. (1995a) 'Riot police thwart animal welfare protest', *The Guardian,* 19 January.

Erlichman, J. (1995b) 'Dover police braced for new livestock protests', *The Guardian,* 14 April.

Feagin, J. R. and Hahn, H. (1973) *Ghetto Revolts: the politics of violence in American cities,* New York: Macmillan.

Field, S. and Southgate, P. (1982) *Public Disorder: a review and a study in one inner-city area,* London: HMSO.

Fielding, N. (1991) *The Police and Social Conflict: rhetoric and reality,* London: Athlone Press.

Fine, B and Millar, R. (eds) (1985) *Policing the Miners' Strike,* London: Lawrence and Wishart.

Foster, J. (1992) 'Scots highlight cycle of estate violence', *The Independent,* 8 July.

Fowler, N. (1979) *After the Riots: the police in Europe,* London: Davis-Poynter.

Gaffrey, J. (1987) *Interpretation of Violence: the Handsworth riots of 1985,* Warwick: University of Warwick, Centre for Research in Ethnic Relations.

Gaskell, G. and Benewick, R. (eds) (1987) *The Crowd in Contemporary Britain,* London: Sage.

Geary, R. (1985) *Policing Industrial Disputes: 1893 to 1985,* Cambridge: Cambridge University Press.

Gennard, J. (1984) 'The implications of the Messenger Newspaper Group dispute', *Industrial Relations Journal*, 15.

Gifford, Lord (1986) *Report of the Independent Inquiry into Disturbances of October 1985 at the Broadwater Farm estate, Tottenham,* London: Karia Press.

Goodman, G. (1985) *The Miners' Strike,* London: Pluto Press.

Gordon, P. (1987) 'Towards the local police state', in P. Scraton (ed.) *Law, Order and the Authoritarian State*, Milton Keynes: Open University Press.

Gould, R.W. and Waldren, M.J. (1986) *London's Armed Police: 1829 to the present*, London: Arms and Armour Press.

Greater London Council (1985) *Public Order Plans: the threat to democratic rights,* London: GLC Publications.

Greater London Council Police Committee (1986) *Policing London: collected reports,* London: GLC Police Committee Support Unit Publications.

Green, P. (1990) *The Enemy Without: policing and class consciousness in the miners' strike,* Buckingham: Open University Press.

Hain, P. (1986) *Political Strikes: the state and trade unionism in Britain,* New York: Viking.

Hall, S. (1980) *Drifting into a Law and Order Society,* London: Cobden Trust.

Hall, S., Critcher, C., Jefferson, T., Clarke, J. and Roberts, B. (1978) *Policing the Crisis: mugging, the state and law and order,* Basingstoke: Macmillan.

Hall, S. and Jefferson, T. (eds) (1976) *Resistance Through Rituals,* London: Hutchinson.

Halloran, J., Elliot, P. and Murdock, G. (1970) *Demonstrations and Communication: a case study,* Harmondsworth: Penguin Books.

Hillyard, P. and Percy-Smith, J. (1988) *The Coercive State*, London: Fontana.

Hobsbawm, E.J. (1959 and 1971) *Primitive Rebels*, Manchester: Manchester University Press.

Hobsbawm, E.J. and Rudé, G. (1970) *Captain Swing*, Harmondsworth: Penguin.

Holdaway, S. (ed.) (1979) *The British Police,* London: Edward Arnold.

Holdaway, S. (1983) *Inside the British Police,* Oxford: Blackwell.

Home Office (1980) *Review of Arrangements for Handling Spontaneous Disorder,* London: HMSO.

Home Office (1983) *Manpower, Effectiveness and Efficiency in the Police Service*, Circular 114/83, London: Home Office.

Home Office (1985) *Review of Public Order Law*, London: HMSO.

Hornsby, M. (1995) 'Woman is killed in veal lorry protest', *The Times*, 2 February.

Hytner, B. (1981) *Report of the Moss Side Enquiry Panel to the Leader of the Greater Manchester Council*, Manchester: Greater Manchester Council.

Jefferson, T. (1990) *The Case Against Paramilitary Policing*, Milton Keynes: Open University Press.

Jefferson, T. and Grimshaw, R. (1984) *Controlling the Constable: police accountability in England and Wales*, London: Frederick Muller.

Johnston, A. (1992a) 'Fury at police simmers on estate', *The Guardian*, 15 May.

Johnston, A. (1992b) 'Joyriders deaths were "last straw" for riot area', *The Guardian*, 18 July.

Johnston, L. (1993) 'Privatisation and protection: spacial and sectoral ideologies in British policing and crime prevention', *The Modern Law Review*, 56 (6): 771-792.

Joshua, H. and Wallace, T. (1983) *To Ride the Storm: the 1980 Bristol 'riot' and the state*, London: Heinemann.

Judge, A. J. (1994) *An Analysis of the Implications for Policing in England and Wales of the Contemporary Profile of New Age Travellers* (unpublished MA dissertation), Leicester: University of Leicester, CSPO.

Kahn, P., Lewis, M., Livock, R. and Wiles, P. (1983) *Picketing Industrial Disputes, Tactics and the Law*, London: Routledge and Kegan Paul.

Katz, I. (1991) 'Vendetta triggered Cardiff "bread riots"', *The Guardian*, 3 September.

Katz, I. (1992) 'Youths and police clash after bikers crackdown', *The Guardian*, 14 May.

Kaye, T. (1984) *A Village at War: a report to the Yorkshire area of the National Union of Mineworkers concerning the policy of policing Fitzwilliam, West Yorkshire, on Monday, 9th July, 1984* (unpublished report), Warwick: Warwick University.

Keith, M. (1993) *Race, Riots and Policing: lore and disorder in a multi-racist society*, London: University of London Press.

Kelly, J. (1981) 'Steel: an irreversible decline?', *Marxism Today*, June.

Kent County Constabulary (1992) *No. 5 Region Command Band Public Order Training: course prospectus 1993/1994*, Maidstone: KCC.

Kettle, M. and Hodges, L. (1982) *Uprising: the police, the people and the riots in Britain's cities,* London: Pan Books.

King, M. (1992) 'Authority and the mob', *The Times Higher Education Supplement*, 20 March: 22.

King, M. (1995) 'Police co-operation and border controls in a "new" Europe', in L. Shelley and J. Vigh (eds) *Social Changes, Crime and the Police*, Chur: Harwood: 149-162.

King, M. and Brearley, N. (1993) *Changing Strategies in Policing Public Order: the British experience: a report submitted to the Canadian Police College, Ottawa*, Leicester: CSPO.

Kinsey, R., Lea, J. and Young, J. (1986) *Losing the Fight against Crime,* Oxford: Basil Blackwell.

Kitson, F. (1971) *Low Intensity Operations: subversion, insurgency, peace keeping,* London: Faber and Faber.

Lavalette, M. and Mooney, G. (1990) 'Undermining the "north-south divide"? Fighting the poll tax in Scotland, England and Wales', *Critical Social Policy,* 10 (2).

Law Commission (1982) *Offences Against Public Order,* Working Paper 82, London: HMSO.

Le Bon, G. (1960 first published 1895) *The Crowd: a study of the popular mind*, New York: Viking Press.

Lea, J. and Young, J. (1984) *What's to be Done About Law and Order?*, Harmondsworth: Penguin.

Leigh, D.L. (1976a) '"Times" man surrounded, robbed', *The Times*, 31 August.

Leigh, D.L. (1976b) 'Confrontation recalls 1958 riots', *The Times*, 31 August.

Lewis, J. (1975) 'A study of the Kent State incident using Smelser's theory of collective behavior', in R.R.Evans (ed.) *Readings in Collective Behavior*, Chicago: Rand-McNally.

Lewis, J. (1986) 'A protocol for the comparative analysis of sports crowd violence', *International Journal of Mass Emergencies and Disasters,* 4 (2): 211-227.

Lewis, J. (1992) 'Theories of the crowd: some cross-cultural perspectives', in: *Lessons Learned from Crowd Related Disasters,* Easingwold Papers No. 4, Home Office Emergency Planning College.

Litton, I. and Potter, J. (1985) 'Social representations in the ordinary explanation of a "riot"', *European Journal of Social Psychology*, 15.

London Strategic Policy Unit (1987) *Policing Wapping: an account of the dispute 1986/ 7*, London: LSPU.

Loveday, B. (1992) 'Reading the runes: the future structure of police forces', *Public Money and Management*, 12 (1).

Loveday, B. (1995) 'Reforming the police: from local service to state police?', *The Political Quarterly*, 66 (2): 141-156.

McCabe, S. and Wallington, P. (1988) *The Police, Public Order and Civil Liberties,* London: Routledge.

McPhail, C. (1991) *The Myth of the Madding Crowd,* New York: Aldine de Gruyter.

Mann, L. (1986) 'Social influence perspective on crowd behavior', *International Journal of Mass Emergencies and Disasters*, 4 (2):171-193.

Manwaring-White, S. (1983) *The Policing Revolution: police technology, democracy and liberty,* Brighton: Harvester Press.

Mark, R. (1977) *Policing a Perplexed Society*, London: Allen and Unwin.

Marnoch, A. (1992) 'Brixton SW9: post-conflict policing', in T. Marshall (ed.) *Community Disorders and Policing: conflict management in action,* London: Whiting and Birch: 93-100.

Marsh, P. (1978) *Aggro: the illusion of violence,* London: J.M. Dent.

Marsh, P., Rosser, E. and Harre, R. (1978) *The Rules of Disorder,* London: Routledge and Kegan Paul.

Marshall, T. (ed.) (1992) *Community Disorders and Policing: conflict management in action* , London: Whiting and Birch.

Marx, G. (1988) *Undercover: police surveillance in America*, Berkeley: California UP.

Mason, G. (1986) 'Duty at fortress Wapping', *Police Review*, December: 2448-2449.

Metropolitan Police (1986) *Metropolitan Police Public Order Review: civil disturbances 1981-1985,* London: Metropolitan Police.

Metropolitan Police (1991) *Trafalgar Square Riot Debriefing, Saturday, 31 March, 1990,* London: Metropolitan Police.

Morgan, R. (1990) *Policing by Consent: current thinking on police accountability in Great Britain,* Social Science Research Centre Occasional Paper 1, Hong Kong: University of Hong Kong, Department of Sociology.

Morgan, R. and Smith, D.J. (eds) (1989) *Coming to Terms with Policing: perspectives on policy,* London: Routledge.

Morrell, W.J. (1992) *The Management of Civil Disorder in the Metropolitan Police* (unpublished MA dissertation), Leicester: University of Leicester, CSPO.

Moscovici, S. (1985) *The Age of the Crowd,* Cambridge: Cambridge University Press.

Myers, P. (1991) 'Police "stood by" as shops looted', *The Guardian*, 4 September.

National Council for Civil Liberties (1968) *Report on the Demonstration in Grosvenor Square, London, on March 17, 1968* (unpublished paper), London: NCCL.

National Council for Civil Liberties (1980) *Southall 23 April 1979: the report of the unofficial committee of enquiry,* Nottingham: Russell Press.

National Council for Civil Liberties (1984) *Civil Liberties and the Miners' Dispute,* London: NCCL.

National Council for Civil Liberties (1986) *No Way in Wapping: the effect of the policing of the News International dispute on Wapping residents,* London: NCCL.

National Council for Civil Liberties (1987) *Public Order Act 1986*, Briefing No.6, London: NCCL.

Nelken, D. (ed.) (1994) *The Futures of Criminology,* London: Sage.

New Society (1981) *Race and Riots '81: a New Society social studies reader,* London: New Society.

New Statesman and Society (1995) *Taking Liberties: civil liberties and the Criminal Justice Act*, London: New Statesman and Society.

Newburn, T. (1995) *Crime and Criminal Justice Policy*, London: Longmans.

Newman, K. (1986) *Metropolitan Police Public Order Review: civil disturbances 1981-1985,* London: Metropolitan Police [also listed as Metropolitan Police (1986)].

Northam, G. (1988) *Shooting in the Dark: riot police in Britain,* London: Faber and Faber.

Norton, P. (ed.) (1984) *Law and Order and British Politics,* Aldershot: Gower.

Ogle, D. (ed.) (1991) *Strategic Planning for Police*, Ottawa: Canadian Police College.

Oldfield, S. (1992) 'Culture clash behind flames of hate', *The Daily Mail*, 24 July.

O'Neill, J. (1995) *The Poverty of Postmodernism*, London: Routledge.

Owusu, K. and Ross, J. (1988) *Behind the Masquerade: the story of the Notting Hill carnival,* London: Edgeware Arts Media Group.

Pearson, G. (1983) *Hooligan: a history of respectable fears,* Basingstoke: Macmillan.

Pearson, G., Blagg, H., Smith, D., Sampson, A. and Stubbs, P. (1992) 'Crime, community and conflict: the multi-agency approach', in D. Downes (ed.) *Unravelling Criminal Justice: eleven British studies,* Basingstoke: Macmillan: 46-72.

Pendlebury, R. (1993) '3,000 lay siege to police station', *The Daily Mail*, 1 February.

Police Monitoring and Research Group (1987) *Policing Wapping: an account of the dispute 1986/7*, Briefing Paper No. 3, London: London Strategic Policy Unit.

Potter, J. and Reicher, S. (1987) 'Discourses of community and conflict: the organisation of social categories in accounts of a "riot"', *British Journal of Social Psychology,* 25.

Potter, J. and Wetherell, M. (1987) *Discourse and Social Psychology: beyond attitudes and behaviour,* London: Sage.

Pryce, K. (1979) *Endless Pressure: a study of West Indian life-styles in Bristol,* Harmondsworth: Penguin.

Ratcliffe, P. (1981) *Racism and Reaction: a profile of Handsworth,* London: Routledge and Kegan Paul.

Reicher, S.D. (1984) 'The St Paul's riot: an exploration of the limits of crowd action in terms of a social identity model', *European Journal of Social Psychology,* 14.

Reicher, S.D. and Potter, J. (1985) 'Psychological theory as intergroup perspective: a comparative analysis of "scientific" and "lay" accounts of crowd events', *Human Relations,* 30.

Reiner, R. (1991) *Chief Constables: bobbies, bosses or bureaucrats?*, Oxford: Oxford UP.

Reiner, R. (1992a) *The Politics of the Police,* London: Harvester Wheatsheaf.

Reiner, R. (1992b) 'Policing a postmodern society', *The Modern Law Review,* 55 (6): 761-781.

Rex, J. (1982) *1981 Urban Riots in Britain,* Leicester: Leicester University.

Roach, J. and Thomaneck, J. (eds) (1985) *Police and Public Order in Europe,* London: Croom Helm.

Roach, L. (1992) 'The Notting Hill carnival: an exercise in conflict resolution', in T. Marshall (ed.) *Community Disorders and Policing: conflict management in action,* London: Whiting and Birch: 101-106.

Rogally, J. (1977) *Grunwick,* Harmondsworth: Penguin Books.

Rollo, J. (1980) 'The Special Patrol Group', in P. Hain (ed.), M. Kettle, D. Campbell, and J. Rollo *Policing the Police*, vol. 2, London: Calder: 153-208.

Rosnow, R.L (1980) 'Psychology of rumours reconsidered', *Psychological Bulletin,* 87.

Rossi, P.H., Berk, R.A. and Edson, B.K. (1974) *The Roots of Urban Discontent: public policy, municipal institutions and the ghetto,* New York: John Wiley.

Rowe, M. (1994) *Race Riots in Twentieth Century Britain,* Occasional Paper in Crime, Order and Policing, Leicester: CSPO.

Rowe, P. and Whelan, C. (eds) (1985) *Military Intervention in Democratic Societies,* London: Croom Helm.

RSPCA (Royal Society for the Prevention of Cruelty to Animals) (1995) 'Kept in the dark', *The Guardian,* 24 January.

Rudé, G. (1964) *The Crowd in History,* New York: John Wiley.

Rudé, G. (1980) *Ideology and Popular Protest*, London: Lawrence and Wishart.

Sadler, S. (1992) 'Crowd control: are there alternatives to violence?' in: T. Marshall (ed.) *Community Disorders and Policing: conflict management in action,* London: Whiting and Birch: 107-128.

Saunders, M. (1992) *Riot Control Technology: a consideration as to the extent to which its use increases the level of violence by both police and dissenters in a riotous situation* (unpublished MA dissertation), Leicester: University of Leicester, CSPO.

Saunders, P. (1981) *Social Theory and the Urban Question,* London: Hutchinson.

Scarman, L. (1974) *Report of Inquiry into the Red Lion Square Disorders of 15 June 1974,* Cmnd 5915, London: HMSO.

Scarman, L. (1986) *The Scarman Report: The Brixton Disorders, 10-12 April 1981,* Harmondsworth: Penguin.

Scraton, P. (1985) *The State of the Police,* London: Pluto Press.

Sears, D.O. and McConahay, J.B. (1973) *The Politics of Violence: the new urban blacks and the Watts riots,* Boston: Houghton Mifflin.

Shapland, J. and Vagg, J. (1988) *Policing by the Public,* London: Routledge.

Sheehy, P. (1993) *Report of the Inquiry into Police Responsibilities and Rewards,* CM2280, II, London: HMSO.

Sherr, A. (1989) *Freedom of Protest, Public Order and the Law,* Oxford: Blackwell.

Silverman, J. (1986) *An Independent Inquiry into the Handsworth/Lozells Riots: 9, 10, 11 September 1985,* Birmingham: City of Birmingham.

Smelser, N.J. (1962) *Theory of Collective Behaviour*, New York: Free Press.

Smith, D.J. and Gray, J. (1985) *Police and the People in London (The PSI Report),* London: Gower.

Solomos, J. (1986) *Riots, Urban Protest and Social Policy: the interplay of reform and social control,* Warwick: University of Warwick, CRER.

South Yorkshire Police (1980) *The National Steel Strike 1980* (unpublished report), South Yorkshire Police.

South Yorkshire Police (1985) *Policing the Coal Industry Dispute in South Yorkshire,* South Yorkshire Police.

Southgate, P. (1982a) 'The disturbances of July 1981 in Handsworth, Birmingham: a survey of the views and experiences of male residents', paper presented at the 20th International Congress of Applied Psychology, Edinburgh University, July.

Southgate, P. (1982b) *Police Probationer Training in Race Relations,* London: Home Office.

State Research Bulletin (1980) 'Policing the Eighties: the iron fist', 3 (19): 146-168.

Stenning, P. (1989) 'Private police and public police: toward a definition of the police role', in D.J. Loree (ed.) *Future Issues in Policing: symposium proceedings*, Ottawa: Canadian Police College:169-192.

Stephens, M. (1988) *Policing: the critical issues,* Hemel Hempstead: Wheatsheaf.

Sullivan, T. J. (1977) 'The "critical mass" in crowd behaviour: crowd size, contagion and the evolution of riots', *Humbolt Journal of Social Relations,* 42: 46-59.

Sumner, C. (ed.) (1982) *Crime, Justice and the Mass Media,* Cambridge: University of Cambridge, Institute of Criminology.

Tan, Y.H. (1992) 'Police public order manual protected', *The Independent,* 11 November.

Tendler, S. and Huckerby, M. (1977) 'Carnival is marred by violence', *The Times,* 29 August.

Tendler, S. and Leigh, D. (1976) '250 are hurt as Notting Hill carnival erupts into violence,' *The Times* , 31 August.

Thackrah, J.R. (ed.) (1985) *Contemporary Policing,* London: Sphere Books.

The Times (1977) '139 hurt, 53 arrested after carnival violence', 30 August.

Thompson, E.P. (1968) *The Making of the English Working Class,* Harmondsworth: Penguin.

Thompson, E.P. (1971) 'The moral economy of the English crowd in the eighteenth century', *Past and Present,* 50: 76-136.

Tilly, C. (1979) 'Collective violence in European perspective', in: H.D. Graham and T.R. Gurr (eds) *Violence in America: historical and comparative perspectives,* Beverly Hills: Sage: 83-118.

Townshend, C. (1993) *Making the Peace: public order and public security in modern Britain*, Oxford: Oxford UP.

Travis, A. (1995) 'Police on beat to test CS gas for protection', *The Guardian*, 14 April.

Tuck, M. and Southgate, P. (1981) *Ethnic Minorities, Crime and Policing,* London: Home Office Research Unit.

Tumber, H. (1982) *Television and the Riots,* London: Broadcasting Research Unit.

Uglow, S. (1988) *Policing Liberal Society,* Oxford: Oxford University Press.

Venner, M. (1981) 'The disturbances in Moss Side, Manchester', *New Community,* 9 (3).

Vogler, R. (1991) *Reading the Riot Act: the magistracy, the police and the army in civil disorder,* Milton Keynes: Open University Press.

Waddington, D. (1992) *Contemporary Issues in Public Order: a comparative and historical approach*, London: Routledge.

Waddington, D., Jones, K. and Critcher, C. (1989) *Flashpoints: studies in collective disorder*, London: Routledge.

Waddington, D., Wykes, M. and Critcher, C. (1990) *Split at the Seams?: community, continuity and change after the 1984-5 coal dispute,* Milton Keynes: Open University Press.

Waddington, P.A.J. (1988) *Arming an Unarmed Police: policy and practice in the Metropolitan Police*, London: Police Foundation.

Waddington, P.A.J. (1994) *Liberty and Order: public order policing in a capital city,* London: UCL Press.

Wainwright, M. (1991a) 'Riot puts despair on agenda', *The Guardian*, 11 September.

Wainwright, M. (1991b) 'Hopes buried in ashes of Meadow Well', *The Guardian*, 11 September.

Wainwright, M. (1995) 'Street kick-about sparked a chain reaction of anger', *The Guardian*, 12 June.

Warpole, K. (1979) 'Death in Southall', *New Society*, May.

Weir, S. (1977) 'The police and the pickets', *New Society*, June.

Welsh Campaign for Civil and Political Liberties and National Union of Mineworkers (South Wales Area) (1985) *Striking Back*, Cardiff: WCCPL and NUM (South Wales Area).

Whitaker, B. (1979) *The Police in Society*, London: Eyre Methuen.

Williams, D. (1967) *Keeping the Peace: the police and public order*, London: Hutchinson.

Williams, E. (1994) 'Investigating major disorder', *Policing*, 10 (2): 134-140.

Wilsher, P., Macintyre, D. and Jones, M. (eds) (1985) *Strike: a battle of ideologies; Thatcher, Scargill and the miners*, Sevenoaks: Coronet.

Worsnip, A. (1995) 'Police trial CS gas', *Leicester Herald and Post*, 17 May.

Wren-Lewis, J. (1981/82) 'The story of a riot: the television coverage of civil unrest in 1981', *Screen Education*, 40.

Wright, S. (1978) 'New police technologies: an exploration of the social implications and unforeseen impacts of some recent developments', *Journal of Peace Studies*, XV (4): 305-322.

Zajonic, R. (1965) 'Social facilitation', *Science*, 149: 269-274.

Zimbardo, P.G. (1969) 'The human choice: individuation, reason and order versus deindividuation, impulse and chaos', in W.J. Arnold and D. Levine (eds) *Nebraska Symposium on Motivation* 17, Lincoln: University of Nebraska Press.

Index

I

J

K

L

M

N